IØØ51167

Felix Publishing 2018
www.felixpublishing.com.au
email: info@felixpublishing.com
Print copies available from publisher.

**Beyond Planet Earth**
Part of the Series **ADVENTURES in EARTH SCIENCE**
Other books in the series include:
Exploration Science (Field Geology and Mapping)
Riches from the Earth (Minerals, Mining & Energy)
Changing the Surface (Erosion and Landscapes)
Rocks – Building the Earth
Fossils – Life in the Rocks
A Dangerous Planet (Earth Hazards)
Through Sea and Sky (Oceanography & Meteorology)

2018  digital book release
ISBN: 978-0-9946432-3-0
2018 Second Print Edition
ISBN: 978-0-9946432-4-7
Author:  Dr P.T.Scott

All illustrations, photographs and videos by the Author unless otherwise stated.Cover photograph: Star cluster and gas in the Carina Nebula courtesy of NASA

Registration:
Thorpe-Bowker +61 3 8517 8342
email: bowkerlink@thorpe.com.au

FELIX

To my grandchildren who are
yet to find their own adventures.

# Beyond

# Planet Earth

## Dr. Peter T. Scott

First released 2017
All rights reserved Felix Publishing

FELIX

## About the Author

Dr. Peter Scott is an award-winning teacher of Earth Science of over forty years' experience in both secondary and tertiary education. he has served on numerous committees in both education and science in several Australian states and was head of the syllabus committee in Earth Science for the State of Queensland. He established science teaching method programs at the University of New England (Australia), lecturing in this program and was an active member of several scientific societies. He holds a Bachelors' Degree, two Masters' Degrees and a Doctorate through extensive field work and research. A keen amateur astronomer, he also taught this subject in secondary schools and was director of the school's observatory.

Dr. Scott at his school observatory, Brisbane, Australia

# Table of Contents

# Chapter 1: Planet Earth

## 1.1 Introduction

From the most distant time in recorded history, humankind has had a fascination with the stars and the other heavenly bodies. The Sun, Moon and stars have all had a part to play in early civilisations, whether it be in religious ceremonies worshiping these recurring objects, or in recording the seasons for planting and harvest, ancient peoples have studied the night sky. **Astronomy** (from the Greek *astron* for the stars and - *omy* for study of) is one of the oldest of the sciences and much of what is taught today owes a considerable debt to the early civilisations of the Babylonians, Egyptians, Greeks, Indians, Chinese, Mayans, Incas, Arabs and the astrologers of the Middle Ages.

## 1.2 Planet Earth (Symbol - ⊕ A Globe With Main Coordinates)

We live on a small planet which orbits a medium-sized yellow star within the Milky Way Galaxy. The name Earth comes from the Old English word *eortha* for ground. In Greek, the planet was called *Gaia* for land and was the primal mother goddess of the Ancient Greeks and in Latin, she and the planet were referred as *Terra*.

Figure 1.1: The Earth seen from space (Photo: NASA)

The Earth is a **planet**. That is, it reflects light only and orbits a star (our Sun) in an elliptical orbit with its closest position (**perihelion**) of 147,095,000 km and its furthest distance (**aphelion**) of 152,100,000 km. – an average of about 149.6 million kilometres.

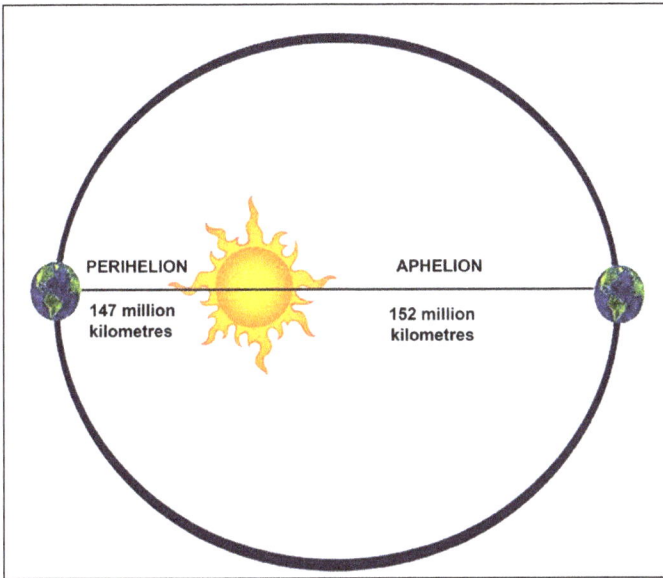

Figure 1.2: The trajectory of the Earth around the Sun

Some of the main facts about our planet Earth are:

- average distance from the sun of 149.6 million km (1 astronomical unit au)

- orbital **period** (year) of 365.26 days

- average orbital velocity of 107200 km/hour

- **sidereal rotation period** (day) is the time for one complete rotation relative to the background stars) of 23 hours 56 minutes 4.100 seconds

- equatorial rotation velocity of 1,674.4 km/hour

- equatorial diameter of 12,756.2 km. and a polar diameter of 12,713.6 km

- axis of rotation is tilted at 23.4393° away from the ecliptic (plane of the orbit of the planets around the Sun)

- surface area of 510,072,000 km$^2$ of which 70.8 % is water and 29.2 % land

- volume of 1.08321×10$^{12}$ km$^3$

- mass of 5.97237×10$^{24}$ kg

- average **density** of 5.514 g/cm$^3$ (i.e. 5.514 times the density of water)

- gravitational acceleration of 9.807 m/s$^2$

- escape velocity or the speed required to break free from the surface, of 40,270 km/hour

- active tectonism from movement of crustal plates giving volcanoes and earthquakes

- magnetic field generated in the liquid outer core by its rotation and free electrons which exist at the high temperature. At the Equator the magnetic-field strength at the surface is 3.05 × 10$^{-5}$ teslas (T). This magnetic field also acts as a shield to the radiation from the solar wind and cosmic radiation from deep space which are trapped by it and form the Van Allen belts which extend from an altitude of about 1,000 to 60,000 kilometres above the surface

- surface temperatures vary from -89.2°C to 56.7 °C with an average of 15 °C

- air pressure at sea level and standard temperature is 101.325 kPa (1013.25 hectopascals ) with composition of:

| | |
|---|---|
| nitrogen | 78.08% |
| oxygen | 20.95% |
| argon | 0.930% argon |
| carbon dioxide | 0.039% carbon dioxide |

<1.0% water vapour (variable) and traces of other gases

- one natural satellite, the moon

## 1.3 The Formation of the Earth and Other Planets

Evidence from radiometric dating of ancient rocks in some of the shield areas and of meteorites, meteors which have struck the Earth's surface, suggests that the Earth formed about 4.54 billion years ago. The **core accretion model**, an updated version of the earlier **nebular hypothesis**, suggests that during the formation of the Solar System, a huge cloud of dust and gas, mostly hydrogen and helium, known as a solar nebula, collapsed under the influence of its own gravitational pull. The force of gravity is one of attraction, and so material will be pulled into a centre containing a concentration of mass. As this inward pull may be slightly off-centre, the whole mass is set spinning, the angular momentum increases as more mass is pulled into the centre and the new system flattens out. As material such as hydrogen gas is forced together, it heats up and at high temperatures; the nuclei or centres of atoms, are forced together by **nuclear fusion** to form helium and a tremendous amount of heat. These fusion reactions can continue to form the other natural elements.

Figure 1.3: Diagram showing the Earth's formation by accretion (Modified after NASA/FUSE/Lynette Cook)

The Sun contained most of the mass formed in the centre of the nebula and the rest was thrown off by the centripetal force of the spinning disk. As the mass which was thrown off was not completely uniform within the spinning disk, small clumps of material began to form at different distances out from the Sun within swirling eddy currents. Smaller particles drew together or accreted, bound by the force of gravity, into larger particles.

The **solar wind**, the intense stream of charged particles emitted by the thermonuclear reactions within the Sun, swept away the lighter elements, such as hydrogen and helium, from the inner regions, leaving only the heavy, rocky materials to create the smaller Inner planets, including the Earth. The Earth's core formed first, with heavy elements colliding and binding together. The new planet grew by **accretion** until its interior was hot enough to melt the heavier metals such as iron, nickel, manganese, iridium and other similar metals. This resulted in the separation and upward movement of the lighter elements such as silicon and aluminium from the metallic core, giving rise to a layered structure and a magnetic field within the nickel-iron liquid outer core.

After a solid crust had formed, there were many eruptions through the surface by the molten material just below the crust, as well as many impacts from space material still spinning outwards. It is thought that the de-gassing of these early volcanoes produced the first atmosphere, including much of the water vapour which later condensed to form the oceans. It is also thought that this volcanic steam was augmented by water introduced by the bombardment of comets, which are essentially large ice balls. The primal atmosphere contained considerable hydrogen, helium, carbon dioxide, nitrogen and methane, but little free oxygen which, being very chemically active had quickly combined with other elements such as silicon, iron and aluminium to form oxides. Most of the lighter gases such as hydrogen, helium and ammonia were driven further out into space by the solar wind. In time and cooling of the planet, condensation of liquid water and the dissolving of carbon dioxide into it allowed the development of massive amounts of oxygen about 2.5 billion years ago, probably because of the rise of the early blue-green algae.

## 1.4 The Seasons

Because of the axial tilt of about $23.5^0$ to the plane of orbit around the Sun, the Sun's rays strike the Earth's surface at different angles at different times of the year giving the planet its seasons.

The summer **solstice** occurs when the rays of the Sun are striking the Earth's surface more directly in the summer months when the Sun is directly overhead at that location. Between June $20^{th}$ and June $22^{nd}$ is the northern summer solstice at mid-summer in the Northern Hemisphere and in the Southern Hemisphere, the summer solstice occurs between December $20^{th}$ and December $23^{rd}$ each year. The slight variation in days is due to the slight wobble of the Earth's axis called its **precession**. On these days, in their respective hemispheres the Sun will appear to be directly overhead at the Tropic of Cancer in the Northern Hemisphere and the Tropic of Capricorn in the Southern Hemisphere. These **tropics** are named because of the constellations of the **zodiac** which is at its **zenith** or its highest point overhead at that time.

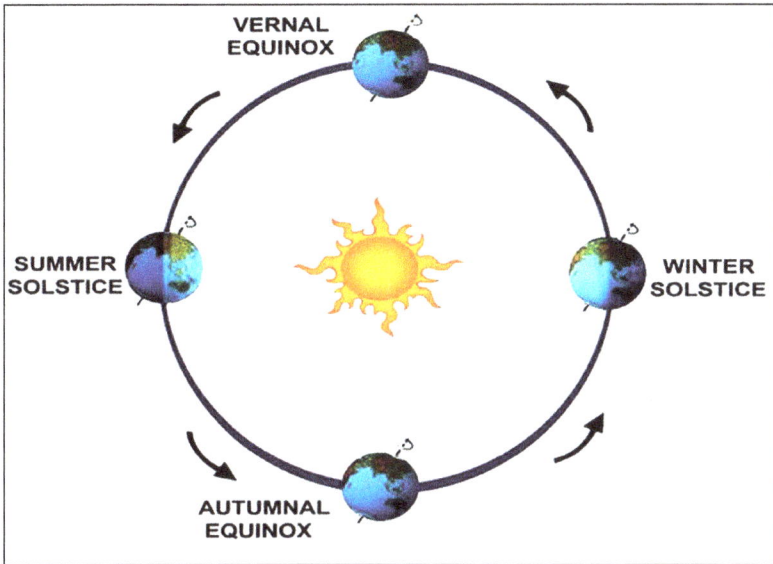

Figure 1.4: Diagram (not to scale) showing the position of a tilted Earth at the four major positions of each Season.

The **equinoxes** occur when the Sun is directly overhead at the Equator and its rays strike the surface at $90^0$, making night and day of equal length all over the planet. Further north and south of the Equator, the Sun now has moved away and is no longer overhead, giving the cooler seasons. These also occurs twice each year, with the autumnal equinox occurring in the Northern Hemisphere marking the beginning of autumn (fall) around the $22^{nd}$-$23^{rd}$ of September and around $20^{th}$ March in the Southern Hemisphere. The vernal equinox marks the beginning of spring, and occurs at these times but is switched for each of the hemispheres so that the vernal equinox in the Northern Hemisphere occurs at the same time as the autumnal equinox in the Southern Hemisphere.

Many ancient peoples including the Babylonians, Egyptians, Celts, Chinese and the Mayans of Central America and Incas further south, all had calendar systems which were used to predict the solstices and equinoxes. This was done for their religious festivals associated with the planting or cultivation times of their crops. The Intihuatana, which means the hitching post of Inti the sun god, in the Incan Quechua language, can be seen at Machu Picchu in Peru, and was an astronomical clock or calendar used by the Incas for their rituals in venerating Inti, the Sun god.

Figure 1.5: The Intihuatana at Machu Picchu, Peru, used by the Incas as a calendar and for their rituals in venerating Inti, the Sun god.

Figure 1.6: Stonehenge on Salisbury Plain, England was a site of Celtic worship and also a solar calendar with stones being aligned to the sunset of the winter solstice and the opposing sunrise of the summer solstice.

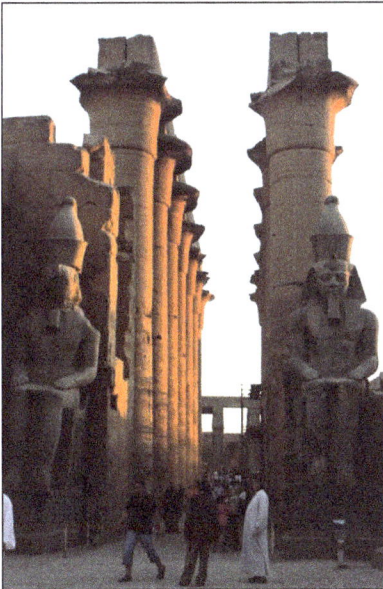

Figure 1.7: The Temple of Amun-Re at Karnak, Luxor, Egypt, was aligned so that on the rising of the midwinter Sun, the whole length of the corridor would be illuminated.

Figure 1.8: El Caracol (the snail), the Mayan observatory-temple at Chichen Itza, Mexico (Photo: Ernest Mettendorf)

## 1.5 The Tides

Tides are the apparent changes in the local levels of the sea due to the gravitational pull of the Moon on the waters of the Earth and the planet's rotation.

Even though the Sun is a very long way from the Earth at 149.6 million kilometres compared to that of the Moon's distance at 384,400 kilometres, it does exert a small gravitational force on the oceans. This causes slight variations in the tides which are noticeable when the Sun and the full moon are in the same line (a situation known as **syzygy** from the Ancient Greek *suzugos* meaning, yoked together) which gives high tides slightly higher than is usual. These are called **spring tides** as the water seems to spring up, not because of the season, and when the Sun and quarter Moon at right angles giving high tides lower than is usual called **neap tides**.

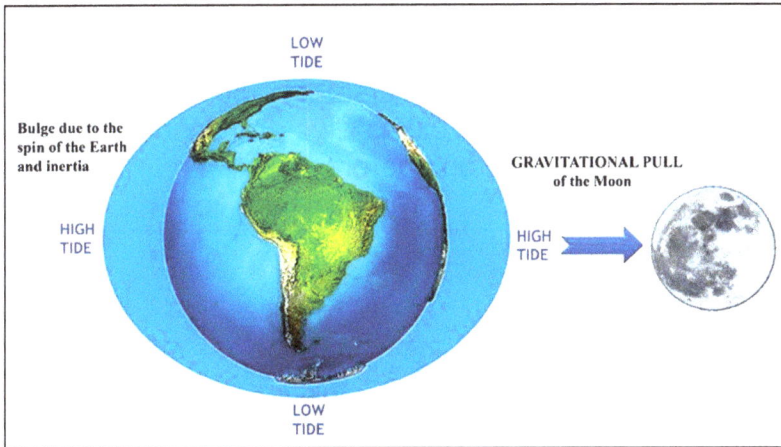

Figure 1.9: Diagram showing how the Moon's gravitational pull produces two high and low tides on opposite sides of the Earth

The height of the tides can also vary during the course of a month because the Moon is not always the same distance from the Earth. As the Moon's orbit brings it in closer proximity to our planet or its **perigee**, from Ancient Greek *peri* for near and *gaea* for Earth, its gravitational forces can increase by almost 50%, and this stronger force leads to higher tides. Likewise, when the Moon is farther away from the Earth or its **apogee**, from

Ancient Greek *apogaion* for near and *gaea* for Earth, the tides are not as high.

Tides usually occur as one high and one low tide in a day and are called **diurnal tides,** but they can also occur as two high waters and two low waters each day and are called **semi-diurnal tides.** These highs and lows do not happen at the same time each day because the Moon takes slightly longer - about 50 minutes more than 24 hours to line up again exactly with the same point on the Earth. Thus the timings of highs and lows are staggered throughout the course of a month, with each tide commencing approximately 24 hours and 50 minutes later than the one before it.

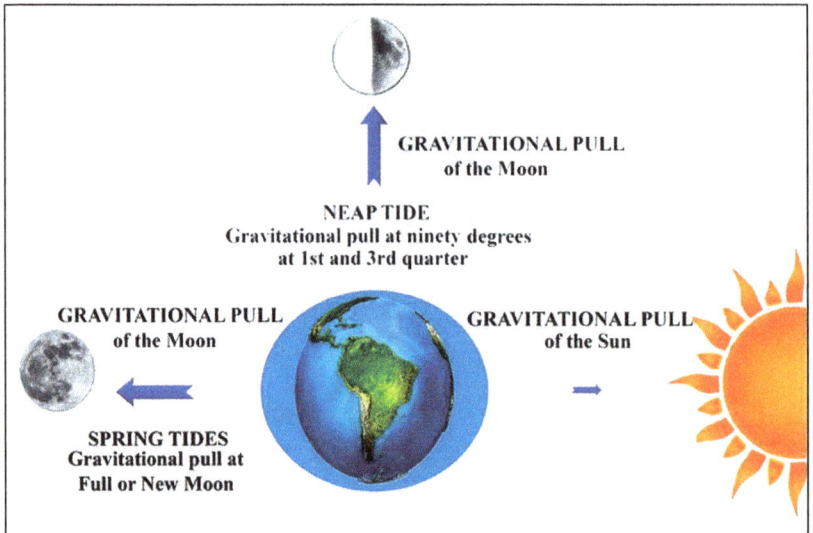

Figure 1.10: Diagram showing spring and neap tides

# Chapter 2: The Moon

## 2.1 The Moon - Satellite of Earth (Symbol: ☾ A Crescent Moon)

The Earth has only one natural **satellite** (a smaller body orbiting a planet) which is called the Moon so named from the Old English Mona which was probably derived from the Proto-Indo-European menses, meaning month. In Ancient Greece, the Moon was named after their moon goddess Selene whom the Romans called Luna.

As a system, the Earth and the Moon orbit each other, but the central point or centre of mass of this orbit or **barycentre**, is actually within the Earth at about 1700 km below its surface and 4,671 from its centre, so the Moon appears to go around the Earth once every 27.3 days (its sidereal period).

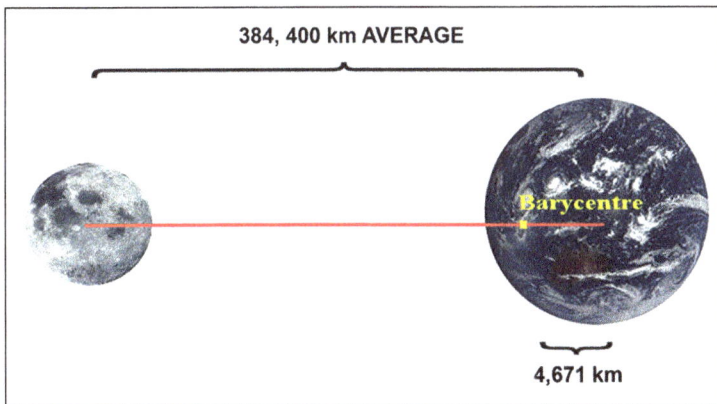

Figure 2.1: Diagram showing the barycentre or centre of mass of the Earth-Moon system

The Moon orbits the Earth at the same time that it rotates which is called **synchronous rotation**. The tidal pull of the Earth causes a locking effect, called **tidal (or gravitational**

**Locking)**, so that the same side of the Moon always faces the Earth. The orbital path is also elliptical with the closest distance, the perigee, being an average of 362,600 km. and the furthest distance, the apogee, being an average of 405,400 km. The Moon's linear distance from Earth is currently increasing at a rate of 3.82 ± 0.07 centimetres per year, but this rate is not constant.

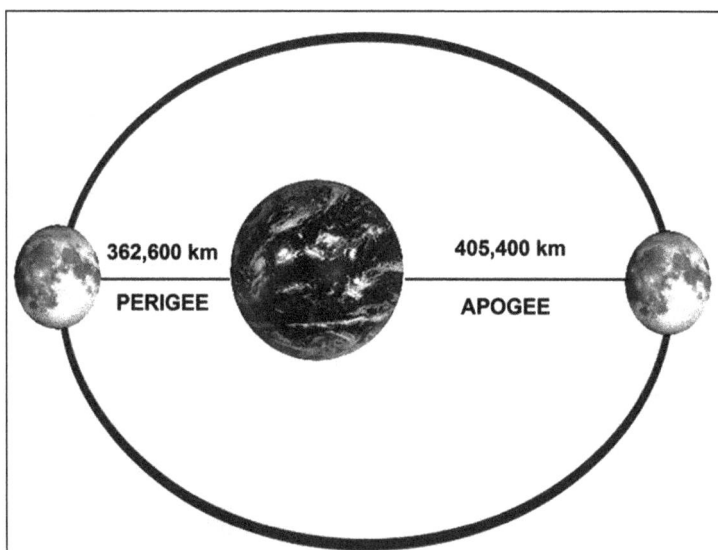

Figure 2.2: Diagram showing the elliptical orbit of the Moon

Some of the main facts about our natural satellite are:

- it is one of the largest natural satellites in the solar system and the largest relative to the size of the planet it orbits (its primary)

- equatorial radius of 1738.14 km  (0.273 that of Earth)

- it is the second-densest satellite in the solar system (after Jupiter's satellite Io) so far known at about at about 3.3464 g/cm$^3$  (0.606 that of Earth)

- gravity is about one sixth (16.7%) that of Earth with a gravitational acceleration of 1.622 m/s$^2$ (compared to Earth's 9.807 m/s$^2$ )

- escape velocity of 142.8 km/hour

- average distance from the Earth of 384,000 km

- orbital period of 27.321582 days

- orbital speed of 1.022 km/s

- orbital tilt around the Earth relative to the ecliptic is about 5°

- axis of rotation is tilted at 5.145° away from the ecliptic

- volume of 2.1958×10$^{10}$ km$^3$  (0.020 Earth volumes)

- mass of 7.3477×10$^{22}$ kg  (0.012300 Earth masses)

- surface temperatures at the equator vary greatly from -173°C (in shadow) to 123°C (in sunlight)

- atmospheric pressure is almost non-existent with a pressure of about $10^{-7}$ Pa (day) to $10^{-10}$ Pa (compared to 101,325 Pa of Earth) being slight traces of the gases hydrogen, helium, argon, neon and radon

- slight traces of water have been found from more recent spectrographic studies from the 2008 Indian unmanned spacecraft, *Chandrayaan-1* suggesting that there may be water ice on the surface, possibly in the permanent shadows of deep craters near the poles

## 2.2 Phases of the Moon

From the Earth, the Moon generally appears to rise in the east (like the Sun) and set in the west because of the rotation of the Earth, but the Moon also appears to change its shape in a regular way, going from one full moon to the next about every 27 days (one moonth in Old English). This apparent change in shape, or **phase,** is due to the relative positions of the Sun, Moon and Earth at different times.

Beginning from a new moon, which is seen as a faint ring or **annulus**, due to sunlight refracting around the edge of the Moon, the shape appears to grow **(waxing)** as a thin crescent moon then into a half-moon at its first quarter and then into a **gibbous** moon and finally into a bright full moon. After this, it slowly diminishes in size **(waning)** to another gibbous moon, half-moon (last or third quarter), a thin crescent and then back to a new moon.

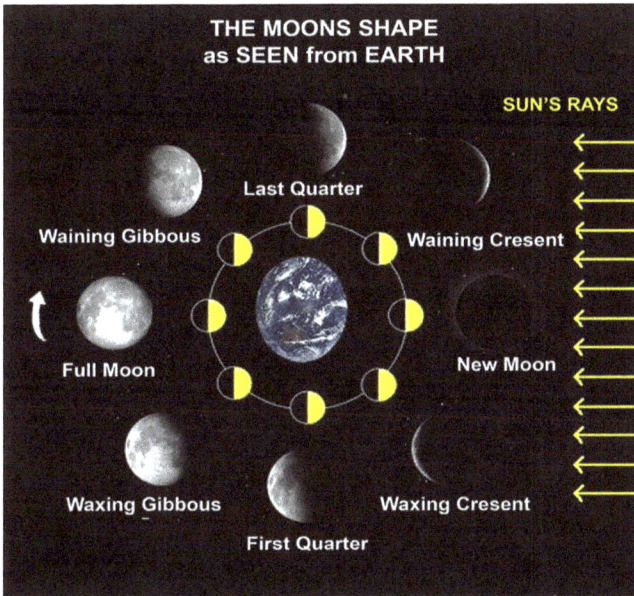

THE MOONS SHAPE
as SEEN from EARTH

SUN'S RAYS

Last Quarter

Waining Gibbous

Waining Cresent

Full Moon

New Moon

Waxing Gibbous

Waxing Cresent

First Quarter

Figure 2.3: Diagram showing the various Phases of the Moon (going clockwise)

## 2.3 Eclipses of the Moon

This occurs at night when the Earth passes between the Moon and the Sun such that the normally full moon will have the shadow of the Earth pass across it obscuring it completely or partially depending upon the relative positions of the three bodies. At a full **eclipse**, the Moon first passes through the **penumbra** or partial shadow, and it appears to go slightly darker, but then the more noticeable effect occurs when the Moon passes through the full shadow or **umbra**. It is because the Moon's orbital axis is tilted at about $5^0$ to the ecliptic, the plane of Earth's orbit around the Sun that this event occurs only about twice a year and not at every full moon. This is an example of a syzygy, a word from the <u>Ancient Greek</u> *suzugos* meaning, yoked together, when the astronomical orientation is perfectly aligned. A partial eclipse occurs when the shadow of the Earth only crosses part of the Moon's surface because of the tilt of the Moon's orbital access at that time.

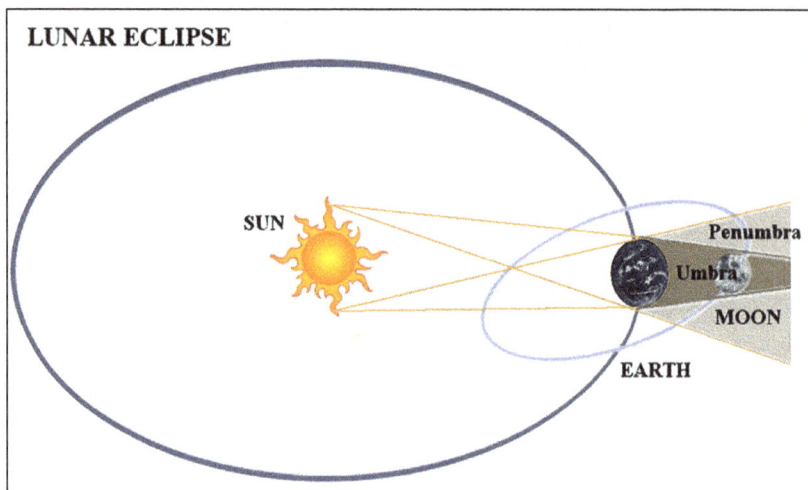

Figure 2.4: Diagram showing a Lunar Eclipse

At total eclipse, the Moon appears a dull red colour because of the refraction or bending of light and the light's **dispersion**, or its scattering into its spectrum of colours by particles in the atmosphere, as it passes through the Earth's atmosphere. Visible light is composed of seven colours, each with their own wavelength – red, orange, yellow, green, blue, indigo and violet. As this light passes through the Earth's atmosphere, the red wavelengths are separated from the others and pass on to the Moon's surface which appears red. This effect on light is called **Rayleigh scattering**, named after **Lord Rayleigh**, (John Strutt, English: 1842-1919) and it also makes the sky on Earth appear red at sunset.

The phases of the Moon and the total lunar eclipse were used by **Aristarchus of Samos** (Greek: 310 – 230 BC) in an attempt to calculate the distances of the Earth to the Sun and the Earth to the Moon. He used direct, naked eye observations of these bodies using simple devices such as the **astrolabe** to measure angles and geometry (Ancient Greek: *geo* for Earth and *metron* for measurement) for his calculations. Aristarchus noticed that the Sun and the Moon appeared to be the same size in the sky but that during an eclipse of the Moon that the Earth cast a curved shadow upon the Moon. He therefore reasoned that the Earth was a sphere and that it came between the Sun and the Moon because the Earth was closer to the Moon than the Sun. He also thought that during the first and third quarters of the lunar phases, the angle between the Sun, Moon and Earth would be at right angles, so he used simple trigonometry to find the relative distances between the Moon and the Earth and the Earth and the Sun.

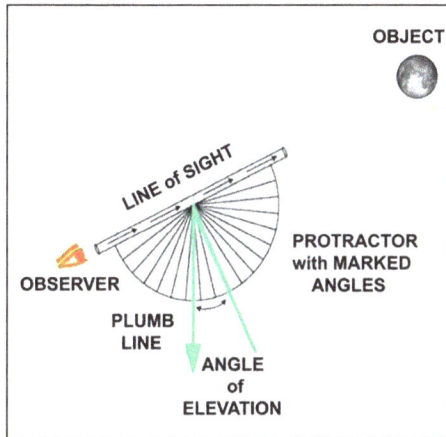

Figure 2.5: Diagram showing how a simple astrolabe would be used to measure the angle of celestial bodies

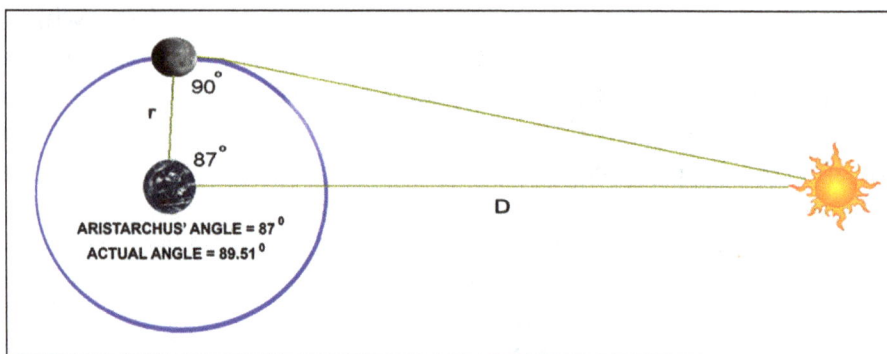

Figure 2.6: Diagram showing the measurement of Aristarchus

He measured the angle to the Moon from his position on Earth and found it to be 87°. Taking the distance to the Moon as r and that to the Sun as D, he was able to find the relative distance to the Sun by the cosine of that angle:

$$\text{Cosine } 87° = r/D$$

$$\text{or} \quad D = r/\cos 87°$$

$$= 19 r \quad \text{(i.e. 19 Earth radii)}$$

This was a good attempt at such a measurement, but the limitations of the early Greek instruments gave an angular value which was much too small. Later measurements using telescopes with very finely-marked angular scales gave the angle at 89.51 ° and a distance to the Sun of about 400 times that of the radius of the Earth.

Aristarchus also attempted to find the, actual distance to the Moon by using the duration of a total lunar eclipse. He estimated, by observation, that the curve of the shadow of the Earth on the Moon would have, if totally projected as a sphere, a radius of a little over twice that of the Earth because it took the about the same time going into the eclipse as it did coming out. He also assumed that the Moon went around a spherical Earth in a

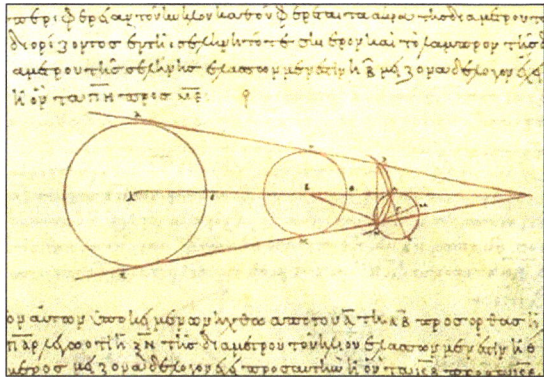

Figure 2.7: The diagram of Aristarchus' calculations from a 10th Century A.D. copy of his part of his work *"On the Sizes and Distances of the Sun and Moon"*.

circular orbit (not quite correct!). He measured the time that the Moon would take to go through this shadow, which was about 3.5 hours, and estimated the distance to the Moon to be about 30 Earth diameters. Whilst his exact method of calculation is not fully known and that his estimate was inaccurate, it did show how early Greek mathematics and observation could be used to calculate distances to objects away from Earth.

## 2.4 Formation of the Moon

There have been several theories as to how the Moon formed in the early days of the Solar System. Our Moon is unusual in that it is very large in comparison with its primary Planet. Some of the major theories include:

- **Co-accretion theory** which suggests that moons can also form at the same time as their parent planet. In this theory, gravity would cause matter in the early Solar System to accrete, or join together, to form the Moon at the same time as the accretion which formed the Earth. Such a Moon would have a very similar composition to the planet, and this would also explain the Moon's present location. However, although Earth and the Moon share much of the same material, the Moon is much less dense than our planet, which would likely not be the case if both started with the same heavy elements at their core.

- **Capture theory** which suggests that the Earth's gravity may have captured a passing body during the early time of the Solar System's formation. The captured moon would then come into a stable orbit around the Earth. This is possibly what has happened with other moons in the Solar System, such as the Martian moons of Phobos and Deimos which have irregular shapes and are considerably smaller than their parent planet. The capture theory would explain any differences in the composition of the Earth and its moon but such orbiters are often oddly shaped, rather than being spherical bodies like the Moon, and their orbits usually do not line up with the ecliptic of their parent planet.

- **Giant impact hypothesis** is the prevailing theory supported by scientific evidence and suggests that the Moon was formed by a glancing impact of a large, perhaps Mars-sized body, termed Theia, impacting the photo-Earth not long after its formation. At impact, much of this body and perhaps part of outer layer of the proto-Earth were thrown into space, accreted together in orbit around the larger Earth forming the Moon. This mode of formation would explain why the Moon is made up predominantly of lighter elements, making it less dense than Earth – the material that formed it came from the crust, while leaving the Earth's core untouched. As the material accreted around what was left of Theia's core, it would have centred near Earth's ecliptic plane, giving the Moon its' current orbital path.

Figure 2.8: Artist's impression of the formation of the Moon by giant impact (Photo: NASA/JPL-Caltech)

In addition to any theory of the Moon's formation, evidence from crater samples from the Apollo Missions suggests that between approximately 4.1 and 3.8 billion years ago, numerous asteroid impacts during what is termed the **late**

**heavy bombardment (LHB)** episode, caused significant changes to the greater surface of the Moon, and by inference, to Earth. Latest reinterpretation of crater statistics however, suggests that the rate of impact on the Moon (and also on Mars) may have been lower than expected, and that the recorded crater population can be explained without any peak in the earliest bombardment of the inner Solar System.

## 2.5 The Surface of the Moon

The surface of the Moon has long been studied by ancient and modern astronomers alike. Because it is only one side of the Moon which faces the Earth, it is the main features of this near side which have been described systematically since the 17th Century. The regolith or surface particles of the Moon consist of fine material broken down by the weathering

Figure 2.9: Gibbous Moon showing some of the main features of the near side. The Terminator, the line between darkness and light is to the right (Photo: NASA).

effects of extreme temperature changes and the constant bombardment of micro-meteorites and solar wind particles.

Figure 2.10: The clear impression of Neil Armstrong's boot (July 20, 1969) shows the fine nature of the Moon's fine regolith (Photo: NASA)

Major surface features of the near side of the Moon facing Earth include:

- **Craters** from the Ancient Greek word for vessel, first used by the Italian astronomer **Galileo Galilei** (1564 – 1642), were once thought to be of volcanic origin but now known as impact craters due to collisions by meteors for smaller craters, and asteroids for larger craters. Many of the older craters were made in the early days of the Moon's existence, but more recent impacts by meteorites have been observed. Some of the older craters have a central spire of rock which is thought to have been produced at impact as the molten rock has rebounded much like dropping an object into water. These new craters have a more sharp-edged rim, are smaller, do not have the flat floors, stepped sides or eroded rims as do the older craters which they sometimes overlap.

- **Maria** (singular: *Mare*. The term was introduced by Johannes Kepler, the German astronomer) which is Latin for seas, as these dark lava fields were thought (in an era of new sea-faring discoveries and wide use of Latin) to be real seas like on Earth. The rock of these maria are similar to Earth basalts, and from Apollo mission sampling and radiometric data appear to have erupted between about 3 and 3.5 billion years ago. The few basaltic eruptions which are seen on the far side of the Moon are much older than those on the near side.

- **Terrae** or highlands are the areas of raised relief which are bright, ancient, rocky areas with a composition richer in lighter minerals such as plagioclase feldspar than the more mafic basalt of the maria. They are more reflective than the maria and, being older, show more impact craters.

- **Rilles** (German for groove) are long, narrow depressions cutting across the maria in some places. Whilst their origin is not fully understood, they are probably collapsed lava tubes caused by volcanic eruptions after the maria had formed, or due to some faulted structure. Lava tubes would form quickly on the surface due to the reduced gravity which would allow them to flow over a long distance, especially with lava of low viscosity. Rilles may be:

  - Sinuous rilles which meander in a curved path and are commonly thought to be the remains of collapsed lava tubes as they usually begin at an extinct volcano.

- Arcuate rilles have a smooth curve and are found on the edges of the dark maria and they may have formed when the lava flows that created the mare cooled and subsided after contraction. These are found all over the Moon.

- Straight rilles which follow long, linear paths and may be narrow grabens (faulted valleys). These can be seen as they pass through craters or mountain ranges.

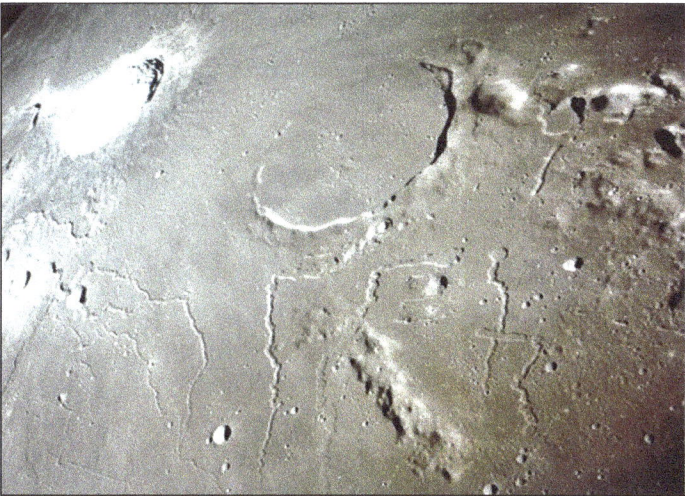

Figure 2.11: Photograph from Apollo 15 showing the older, basalt-flooded Prinz Crater (top, centre), the prominent Aristarchus (top left corner), several smaller recent craters and a number of rilles (lower left and right) (Photo: NASA).

- **Crater rays** are shown as bright, streaks radiating out from some of the more prominent craters. These are thought to have been caused by ejecta thrown out during the impact which formed the crater. The rays can extend for lengths up to several times the diameter of their originating crater and are often accompanied by small secondary craters formed by larger chunks of ejecta.

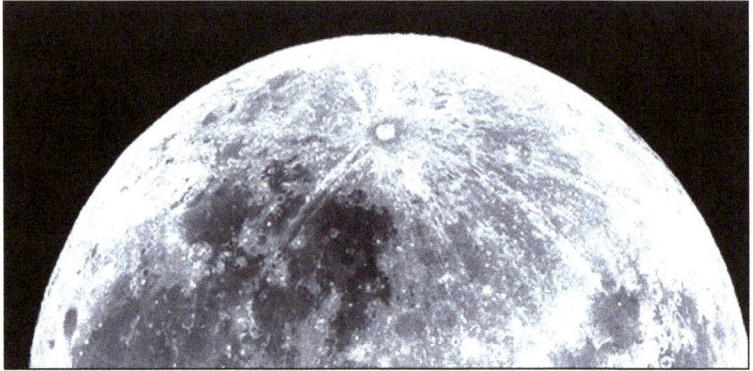

The Moon has a great variety of the names of its surface features. The dark, smooth maria are named for features of the weather or states of mind, such as Mare Imbrium (the Sea of Showers), Mare Tranquillitatis (the Sea of Tranquillity) and Mare Fecunditatis (the Sea of Fecundity). Many of the abundant craters are named for famous scientists, philosophers, mathematicians and explorers such as Copernicus, Tyco, Plato and Kepler. This tradition comes from Giovanni Battista Riccioli, who started it in 1651 and since 1919, assignment of these names has been regulated by the International Astronomical Union.

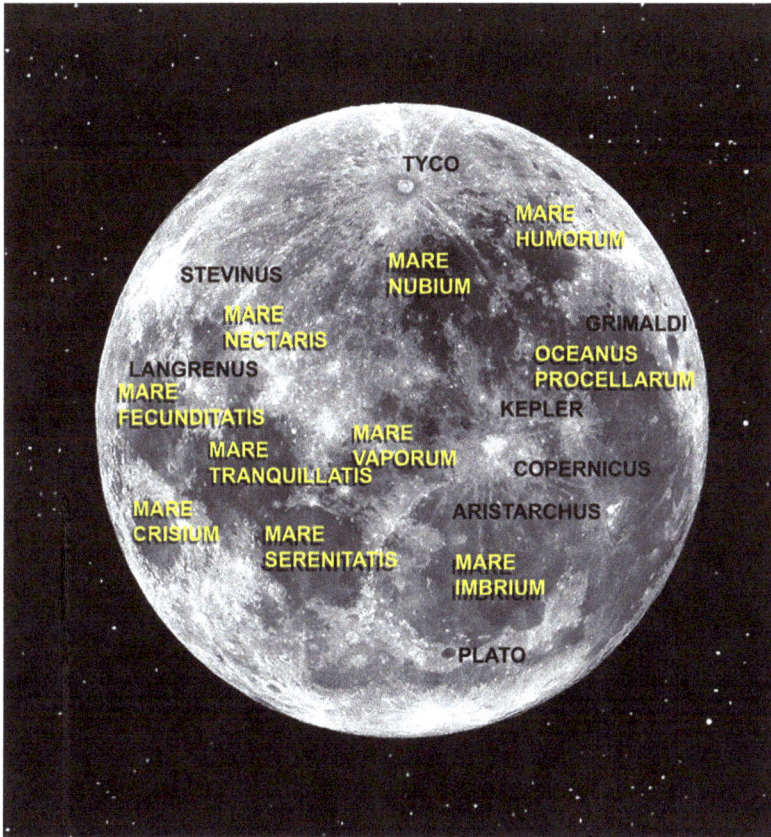

Figure 2.13: Photograph showing some of the major craters (in black) and maria (in yellow) on the near side of the Moon from the Southern Hemisphere of Earth – this image would be inverted if viewed in the Northern Hemisphere (Moon photo: NASA).

The far side of the Moon is often erroneously called the dark side, but it receives as much light as the near side and there is some overlap between the two. However, the two sides have distinctly different appearances, with the near side covered in large maria and the far side having only a few small maria, covering only about 1% of the surface compared to 31.2% on the near side.

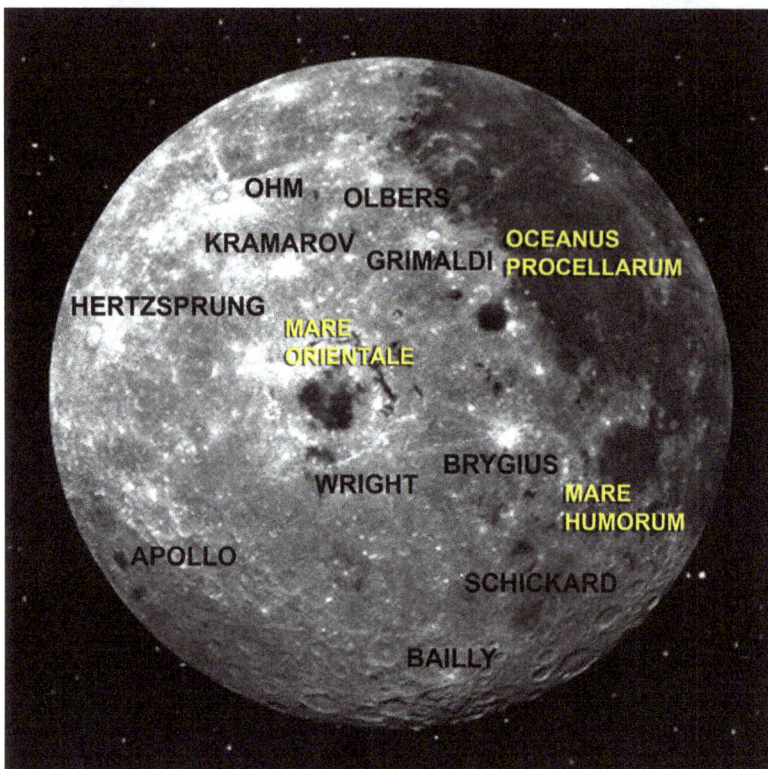

Figure 2.14: Photograph showing the far side of the Moon with the large Mare Orientales, Oceanus Procellarum and Mare Humorum which can also be seen on the near side. Craters also have more modern names. (Photos adapted from NASA).

Data from the Lunar Prospector gamma-ray spectrometer suggests that this difference may be due to a higher concentration of heat-producing chemical elements on the near side, while other factors such as surface elevation and crustal thickness may have limited the spread of basalt lavas. This is also why the far side has more visible craters because there had not been the wide-spread lava fields of the maria to obscure the older craters rather than not being shielded by the Earth. NASA has calculated that the Earth obscures only about 4 square degrees out of 41,000 square degrees of the sky as seen from the Moon which makes shielding of the near side unlikely and that each side of the Moon had received equal numbers of impacts. Most of the craters on the far side were named by the Soviet (now Russian) Academy of Sciences following the successful photography of their Luna-3 Space probe and honour famous astronomers, scientists and science-fiction writers.

Figure 2.15: Photograph of Crater 308 on the far side of the Moon taken by Apollo 11 astronauts on July 1969. Notice the rimmed nature of this crater, its central mountain peaks and the younger craters which have overlapped its rim (Photo:NASA)

## 2.6 The Interior of the Moon

Evidence from studies of the surface rocks and the Moon's density as well as from two seismic stations set up by Apollo missions, suggests that the Moon has a layered structure. The surface rocks are similar to **anorthosites** on Earth which have over 90% plagioclase and the rest are the mafic minerals such as pyroxenes, olivines, magnetite and ilmenite. This crust is very thin, being only about 50 km. on the near side and about 100 km on the far side.

The Moon has a fluid outer core mostly of liquid Iron with a radius of roughly 300 km surrounding a small inner core with a radius of less than 350 km. Its composition is not well understood, but it is probably metallic iron with a small amount of nickel and analyses of the Moon's rotation and the Moon's weak magnetic field suggests that it may be partly molten.

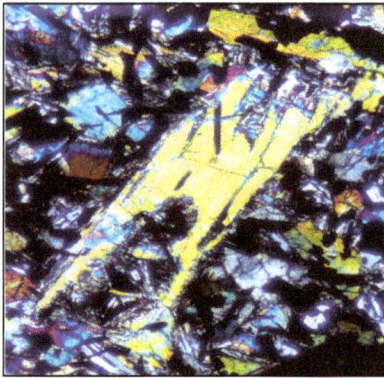

Figures 2.16 & 2.17: A small sample of ilmenite basalt (left - 8cm. across) typical of the mare and brought back by the Apollo missions and a photomicrograph (right with cross polars, about x 25). The large yellow crystal is a pyroxene phenocryst which has been partially resorbed. Other yellows and reds are also pyroxenes whilst the blues and whites are plagioclases and the smaller black shapes are

Below this thin crust is a rocky mantle with a more mafic composition but probably with more iron than in the Earth's mantle. Below this is a partially molten boundary layer with a radius of about 500 km. This is thought to have developed due to fractional crystallization shortly after the Moon's formation 4.5 billion years ago. Just over 700 km deep is a seismic zone from which deep moonquakes have been detected. These may be due to some movement within the mantle by tidal effects with the Earth but are not fully understood.

Figure 2.18: "Buzz" Aldrin from Apollo 11 deploys a seismograph on the Sea of Tranquility, 1969 (Photo: NASA)

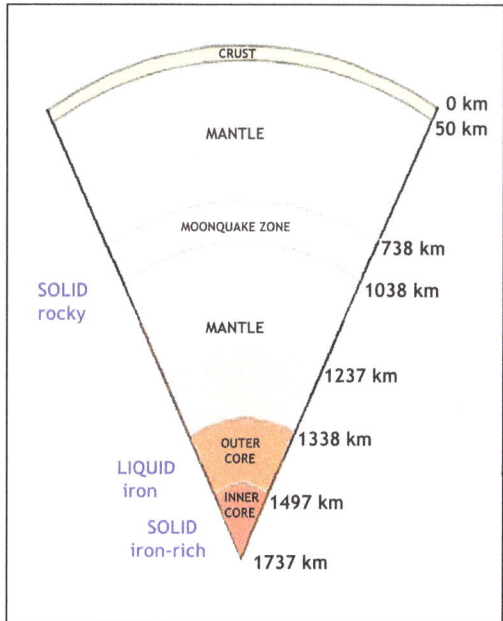

Figure 2.19: Diagram showing a model for the Moon's interior

There are many free Apps. are available online (Search Words: Moon, Lunar Phases, Eclipses, Moon Map – search for Offline Apps.)

## 2.7 From the Earth to the Moon - the Dream

Humankind has long since admired and even worshipped the Moon. The Ancient Greeks worshiped their moon goddess Selene whom the Romans called Luna. Today, **selenology** is the scientific study of the Moon and there are several uses of the term lunar in modern language. In Western Europe, the Celts worshiped Rhiannon, the goddess of fertility as well the Moon.

In Ancient America, Athenesic was a Moon goddess of several North American tribes, whilst further south, the Aztecs worshiped Coyolxauhqui. Meaning golden bells, she was the daughter of the Earth goddess, Coatlicue and the sister of the Sun god, Huitzilopochtli. The Mayans worshiped Awilix and the Incas worshiped Mama Killa or Mother Moon- the sister and wife of Inti, the Sun god. Many ancient peoples worshiped the Moon as a goddess because they identified the monthly cycle of the Moon with that of the menstrual cycle of women; however the Ancient Egyptians worshiped a Moon god, Iah a name which simply meant moon. As well as worship, the Ancient peoples studied and recorded the movement of the Moon and its regular cycle.

When the forces of **Ferdinand** (1452-1516) and **Isabella** (1451-1504) of Aragon and Castile defeated the last Emir of Granada in Moorish Spain in 1492, the scholars who accompanied the army found a great store of literature, including many copies of and references to astronomy texts of the Ancient Greeks which had been lost to the Europeans during the Dark Ages. They also found a wealth of information on the study of astronomy by Arabian, Persians and Indian scholars which now contributed greatly to this ancient science and to the European Renaissance then underway. These lost manuscripts included works from such Astronomers as:

- **Aristarchus** (Greek: 310-230 B.C.) who had put forward the idea that the Sun was actually in the centre of the universe and he also attempted to measure the relative distances between the Earth and the Sun and the Earth and the Moon.

- **Hipparchus** (Greek: 190-120 B.C.) who had measured the distance from the Earth to the Moon to be 29.5 Earth diameters (real value is 30 Earth diameters) and described the precession ( wobble) of the Earth's axis, caused by the gravitational pull of the Sun and Moon.

- **Muḥammad ibn Mūsā al-Khwārizmī** (Persian: 780-850 A.D.) who was the inventor of algebra and made detailed calculations of the positions of the Sun, Moon, and planets and calculated a number of eclipses.

- **Aryabhata** (Indian: 476-550 A.D.) who suggested that reflected sunlight is the cause behind the shining of the Moon and that the Earth rotated on its axis causing the East-West motion of the Moon and stars.

By the year 1492, when **Christopher Columbus** (Italian: 1451-1506) sailed westward to look for the New World, many sea-farers and astronomers believed that the Earth was a sphere, but leaving it was beyond their imagination. There had been many stories from many ancient civilizations which had their heroes traveling away from Earth. In more modern times, some authors with creative imaginations, created science fiction with their heroes leaving the Earth in search of new horizons. In 1638, **Francis Godwin** (English: 1562-1633) wrote *The Man in the Moone* in which his hero flies to the Moon using a cart pulled by geese and in 1657, the trilogy *The Other World: Comical History of the States and Empires of the Moon* written by the Frenchman, **Cyrano de Bergerac** (1619-1655) mentions rockets as the mode of travel for his hero.

More recently, modern science fiction included the works by **Jules Verne** (French: 1828-1905) who wrote *From the Earth to the Moon* in 1865, and its sequel, *Around the Moon* in 1870. In these novels, Verne, the lawyer infatuated with science and the new industrial age in Europe and America, imagined his American heroes being shot to the Moon in a huge projectile fired from the *Columbiad* space gun and then returning to a splash-down in the Pacific Ocean. Whilst Verne did

Figure 2.20: The firing of the Baltimore Gun Club's *Columbiad* in Jules Verne's 1865 novel *From the Earth to the Moon* (Photo: Wikipedia)

try to do some calculations of the physics involved, such transportation was impractical because the length of the gun was inadequate to obtain the escape velocity of Earth of 40,270 km/hour, and that the concussion of such a firing would have killed all on board. Later, in 1901, **H.G. Wells** (British: 1866-1946 ) published his *The First Men in the Moon* in which his hero travels to the Moon using a sphere covered in slats of Cavorite (named after its inventor and Moon traveller) which could "shield gravity" when opened.

Whilst the Chinese had used their invention of gunpowder to propel rockets (named from the Italian *rocchetta* for little fuse) from the 13th Century, and other nations had used more modern versions in the 18th and 19th Centuries for warfare, it was not until the 20th Century that serious research began in rocketry- often by people who had been inspired by Verne and Wells. These included:

**Konstantin Tsiolkovsky** (Russian:1857-1935)
A high school mathematics teacher, published in 1903 his work *The Exploration of Cosmic Space by Means of Reaction Devices*, the first serious scientific work on space travel which advocated the use of liquid hydrogen and oxygen for fuel. His work was essentially unknown outside the Soviet Union, but inside the country it inspired further research, and the basic relationship which describes rocketry – The Tsiolkovsky Rocket Equation was named in his honour. In 1924 he suggested the use of Cosmic Rocket Trains, or multistage rockets to increase range.

Figure 2.21: Konstantin Tsiolkovsky

$$\Delta v = v_{e} \ln \frac{m_0}{m_1}$$

This is Tsiolkovsky's Rocket Equation which describes all rocket propulsion,
where:
$m_0$ is the initial total mass.
$m_1$ is the final total mass.
$v_e$ is the effective exhaust velocity and
$\Delta v$ is the maximum velocity reached.

**Robert H. Goddard** (American: 1882-1945) studied the principles of rocketry and in 1912 suggested that traditional solid-fuel rockets could be improved by: burning the fuel in a small combustion chamber; arranging the rocket as several stages; and that the exhaust speed could be greatly increased beyond the speed of sound by using an hour-glass shaped nozzle (the de Laval nozzle- named after its Swedish inventor, **Gustaf de Laval** (1845-1913). In 1920, Goddard published these ideas in *A Method of Reaching Extreme Altitudes*, which included the view that a solid-fuel rocket could reach the Moon. Further research, however, led Goddard to research into liquid-fuel rockets using the de Laval nozzle which gave a more efficient stream of propulsion gases. On 16 March 1926, Robert Goddard launched the world's first liquid-fueled rocket in Auburn, Massachusetts. Today, the most common propellant combinations used today are liquid oxygen/kerosene and liquid oxygen /liquid hydrogen.

Figure 2.22: Robert H. Goddard

Figure 2.23 Goddard with his rocket (Photo: NASA)

**Hermann Oberth** (German: 1894-1989)
In 1923 after years of study and experimentation with liquid-fueled rockets, Oberth published *The Rocket into Planetary Space.* During the years 1928 and 1929, Oberth worked as a consultant on Fritz Lang's famous science fiction film, *The Woman in the Moon,* which had an enormous influence in popularizing space exploration. Oberth continued to work on both liquid and solid fuel rockets throughout the 1940s and 1950s, both in Germany and then in Italy after World War 2. In 1953, Oberth returned to Germany, where he published his book *Man into Space*, in which is described such advanced ideas as space-based telescopes, space stations, and space suits. Eventually he moved to America in the 1960s and worked for NASA and the Convair Corporation on their rocket programs before retiring in 1962.

Figure 2.24: Hermann Oberth

**Wernher Magnus Maximilian, Freiherr (Baron) von Braun**
(German-American: 1912 – 1977)

In 1929, as an eighteen-year old student, von Braun helped Hermann Oberth with his static firing of his first liquid-fueled rocket motor (named the *Kegeldüse* or cone nozzle) which only ran briefly. In 1930, he joined the *Verein Für Raumschiffahrt* (the VfR or Spaceflight Society) and assisted **Willy Ley** (German-American experimenter, science writer and spaceflight advocate: 1906 – 1969) in his liquid-fueled rocket motor with Hermann Oberth. Wanting to learn more about the science of rocketry, von Braun began post-graduate studies and graduated as a Doctor of Philosophy in Physics in aerospace engineering in 1934. In 1939 he joined the National Socialist German Workers Party (NSDAP, or Nazi party) and during World War 2 he became Technical Director at the German rocket research base at Peenemünde on the North Sea. Here he assisted in the development of the Aggregat (A Series) of rockets based mainly on the plans and journals of Robert Goddard. The A-4 rocket research eventually developed into the infamous missile, the V-2 rocket. At the end of the war in 1945, he was captured by the American Army who took him and many of his colleagues to work for the American rocket program. Here as Director of the

Figure 2.25: Wernher von Braun

Development Operations Division of the Army Ballistic Missile Agency, von Braun with his team developed many of the first American rockets, eventually developing the powerful Saturn V rocket which was used in the Apollo Program to take humankind to the Moon.

The A-4 used a 74% alcohol, as ethanol and water mixture for fuel, and liquid oxygen as the oxidizer. The fuel and oxidizer pumps were driven by a steam turbine with the steam being produced using hydrogen peroxide with a sodium permanganate catalyst. The combustion burner reached a temperature of 2,500°C to 2,700°C. At launch the A-4 propelled itself for up to 65 seconds on its own power reaching an altitude of 80 km when the engine stopped and the rocket continued on a free-fall trajectory.

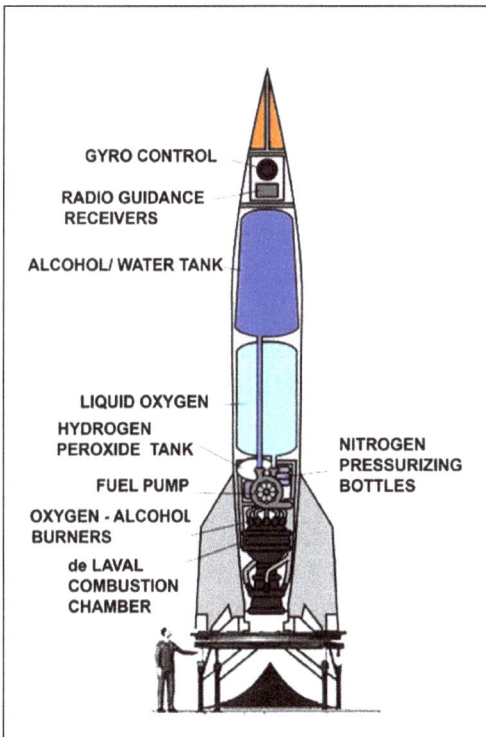

GYRO CONTROL

RADIO GUIDANCE RECEIVERS

ALCOHOL/ WATER TANK

LIQUID OXYGEN

HYDROGEN PEROXIDE TANK

FUEL PUMP

OXYGEN - ALCOHOL BURNERS

de LAVAL COMBUSTION CHAMBER

NITROGEN PRESSURIZING BOTTLES

Figure 2.26: German A2/V2 single-stage rocket cut-away view

## 2.8 From the Earth to the Moon – the Reality

After World War 2, the Americans and the Russians developed their rocket programs both as a military weapon and as exploratory venture into space. After testing with sub-orbital flights, the first object launched into orbit around the Earth was the Russian artificial satellite *Sputnik* 1 (Russian for satellite-1), on October 4th, 1957. It had an orbital velocity of about

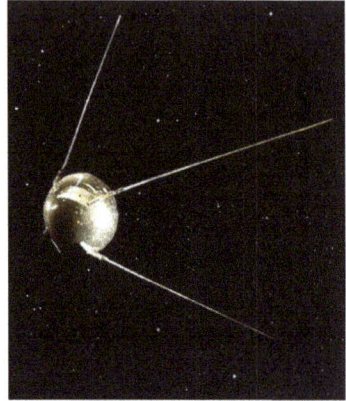

Figure 2.27: An artist's impression of Sputnik 1 in orbit

29,000 kilometres per hour and took only 96.2 minutes for one complete orbit. It contained no scientific instruments except a radio transmitter which sent out a signal at 20.005 and 40.002 megahertz (MHz) which could be detected by amateur radio operators. The signals continued for 21 days until the transmitter batteries ran out on 26 October 1957 and *Sputnik 1* burned up on 4 January 1958. Whilst only a simple device, Sputnik 1 was useful in providing some information from the **ionosphere** and it demonstrated the practicality of putting objects into orbit. The first U.S. satellite, *Explorer* 1, went into orbit later that month on January 31, 1958.

The Russian Luna programme was a series of unmanned missions to the Moon between 1959 and 1976. Fifteen missions were successful with each being either an orbiter or lander. The most notable missions are given in the table below in Table 2.1:

| MISSION | DATE | ACHIEVEMENTS |
|---------|------|--------------|
| LUNA 1 | January 1959 | Missed its intended impact with the Moon and became the first spacecraft to fall into orbit around the Sun |
| LUNA 2 | September 1959 | Successfully hit the Moon's surface, becoming the first man-made object to reach the Moon |
| LUNA 3 | October 1959 | Rounded the Moon and returned the first photographs of its far side. |
| LUNA 9 | February 1966 | Became the first probe to achieve a soft landing on another planetary body. It returned five black and white stereoscopic circular panoramas, which were the first close-up shots of the lunar surface. |
| LUNA 10 | March 1966 | Became the first artificial satellite of the Moon |
| LUNA 15 | July 1969 | Designed to return soil samples from the lunar surface, underwent its mission at the same time as the Apollo 11 mission. Neil Armstrong and "Buzz" Aldrin were already on the lunar surface when Luna 15 began its descent, and the spacecraft crashed into a mountain minutes later. |
| LUNA 16 | September 1970 | Collected samples of lunar soil and returned them to Earth. |
| LUNA 17 | November 1970 | Carried the Lunokhod vehicle, which roamed around on the Moon's terrain. |
| LUNA 20 | February 1972 | Collected samples of lunar soil and returned them to Earth. |
| LUNA 21 | January 1973 | Carried the second Lunokhod vehicle. |
| LUNA 24 | August 1976 | Collect samples of lunar soil and returned them to Earth. The programme returned a total of 0.326 kg of lunar samples |

Figure 2.28: An artist's impression of a Russian Lunokhod robotic rover on the surface of the Moon (Photo:NASA)

On April 12th 1961, Russian Cosmonaut Lieutenant **Yuri Gagarin** (1934 -1968) became the first human to orbit Earth in his spacecraft *кедр* or Cedar, the first of the *Vostok* missions (Russian: *Восток-1* or *East-1*). His single orbit lasted 108 minutes reaching an altitude of 196 kilometres and upon re-entry, Gagarin ejected from *Vostok* at an altitude of 7000 metres and parachuted to Earth. On May 5th, 1961, **Alan Shepard**

Figure 2.29: Yuri Gagarin - first human in space

(1923 – 1998) became the first American to fly into space on board *Freedom 7* which made only a short, sub-orbital flight for 15 minutes. It was not until February 20, 1962, that **John Glenn** (1921 - ) became the first American to orbit Earth aboard *Friendship 7*, circling the globe three times during a flight lasting 4 hours, 55 minutes, and 23 seconds. The race into space had begun.

Online Video 2.1: Yuri Gagarin is launched into space (modified from NASA archives) Go to https://www.youtube.com/watch?v=4CZ48a7kU2Q)

46

A list of the major manned missions into space is given in the table below:

| MISSION (COUNTRY) RUSSIAN AMERICAN | LAUNCH DATE | CRAFT | CREW | MISSION HIGHLIGHTS (MOON MISSIONS HIGHLIGHTED) |
|---|---|---|---|---|
| Vostok-1 to 6 (USSR) | April 12, 1961 until June 16, 1963 | Kedr (Cedar) Orel (Eagle) Sokol(Falcon) Berkut(Golden Eagle) Yastreb (Hawk) Chaika(Seagull) | Gagarin Titov Nikolayev Popovich Bykovsky Tereshkova | Yuri Garagin first human in space. Vostok 3 & 4 flew together. Vostok 5 had 5 days in space and Valentina Tereshkova was the first Woman in space. |
| Mercury- 3,4,6,7 8 & 9 (USA) | May 5, 1961 to May 15, 1963 | Freedom 7 Liberty Bell-7 Friendship 7 Aurora 7 Sigma 7 Faith 7 | Shepard Grissom Glenn Carpenter Schirra Cooper | Shepard was the first American in space and Glenn made the first orbital flight. |
| Voskhod 1 & Voskhod 2 (USSR) | October 12, 1964 March 18, 1965 | Rubin (Ruby) Almaz (Diamond) | Komarov Feoktistov Yegorov Belyavyev Leonov | The first space crew, with one pilot and two passengers and first spacewalk. |
| Gemini 3, 4,5,7,6-A, 8, 9-A, 10,11 & 12 (USA) | March 23, 1965 | Various names | Each had a two-man crew. | They could be steered by its own rockets, had space walks and rendezvous and docking experiments with an Agena target capsule |

47

| | | | | and fellow Gemini craft. |
|---|---|---|---|---|
| Apollo 1, 7, 8, 9, 10, 11, 12, 13, 14, 15, 16, 17 (USA) | January 27, 1967 to December 7, 1972 | Various names for each mission, command and lander modules | 3-man crew with 1 staying on board the Command Module and 2 landing on the Moon's surface | Astronauts of Apollo 1 were killed in a fire. Missions 7-10 went to the Moon but Apollo 11 was the first to land with Astronauts Armstrong & Aldren landing and Collins staying in orbit. Apollo 13 malfunctioned but was able to return to Earth. Apollo 15 used a manned rover. |
| Soyuz Missions (USSR) 66 missions | April 23, 1967 to October 2, 1991. | Various names with the first being Rubin (Ruby) and the last Derbent ( a city in Dagestan) | Many Russian and other Cosmonauts from Soviet-bloc countries and guests from Japan, France, UK and India. | Starting with single Cosmonauts, this program carried out many rendezvous and dock procedures with other craft in the series, setting endurance records for length of time and number of crew. |

48

| Salyut-1, 3 to 7 (excluding two failed attempts) Space Station | 19 April 1971 | No specific names for each mission. Salyut means salute in English. | Crew usually 2 to 4 but later missions had up to 33 visitors, all arriving on Soyez spacecraft | Successful space station experiments, mostly civilian and some military. Salyut 7 had 3,216 days manned by different crew. |
|---|---|---|---|---|
| Skylab Missions 2, 3 & 4 (USA) | May 25, 1973 to November 15, 1973 | One Skylab Space Station with added modules | Conrad Kerwin Weitz<br><br>Bean Garriott Lousma<br><br>Carr Gibson Pogue | The first American laboratory in space, the Skylab spent up to 84 days in space performing experiments and observations. |
| Apollo-Soyuz Test Project (USA-USSR) | July 15, 1975 | Apollo/Soyuz | Leonov Kubasov Stafford Brand Slayton | A joint mission to link the last Apollo spacecraft with the Soyuz 19 spacecraft. On July 17, and the two crews shake hands. |
| The "Space Shuttle" series -various designations (USA) | April 12, 1981to July 8, 2011 | Columbia (27 trips) Challenger (10) Discovery (39) Atlantis (33) Endeavour (25) | From 2 to 8 crew (later missions) | The first winged, reusable spacecraft. Sally Ride of STS-7 first American woman in space. Columbia broke apart on re-entry & Challenger exploded at its 10$^{th}$ |

| | | | | launch. All crews were killed. Assisting in the building of space stations, launching and recovery of satellites, European Space Agency and other experiments |
|---|---|---|---|---|
| Mir Space Station | 19 February 1986 (Core) with several modules (Kvant) added later. Program finished 23 March 2001 | Mir with Kvant modules | Many Cosmonauts with guests from many European countries, American Astronauts and Japan. | Many Space Station and other experiments and docking practice between Russian and American craft. Allowed to burn up in Earth's atmosphere |
| International Space Station (ISS) | 20 November 1998 to present | No specific name - space station and many added modules serviced by Russian, American and European spacecraft | A large number of crew and visiting scientists from many countries. Usually no more than 6 crew at a time. | Space Station experiments and a wide range of scientific experiments including those providing habitat data for longer space exploration. |

Table 2.2: Some early manned space missions (Russian is coloured orange, American in blue and joint missions in green)

50

## 2.9 The Apollo Mission to the Moon

The first Apollo mission was launched aboard a Saturn V launch vehicle - the V designation is for the five powerful F-1 engines that powered the first stage of the rocket. The Saturn V still remains the largest and most powerful U.S. expendable launch vehicle ever launched and the payload of the Apollo spacecraft was probably at its maximum limit. The Saturn V launch vehicle stood 110 metres tall and consisted of three stages:

- First Stage (S-IC) weighed 2,000 metric tons and with its five F-1 engines producing nearly 34,000 kilonewtons of thrust(1kN = 1000 N each being the force required to push 1 kilogram to an acceleration of 1 metre/second squared) . These engines were powerful enough to accelerate the rocket to the escape velocity of the Earth. The first stage engines lasted for about 2.5 minutes taking the vehicle to an altitude of 60 kilometres. It then separated and burnt up in the Earth's atmosphere.

- Second Stage (S-II) weighed 480 metric tons and had five J-2 engines giving a thrust of 4,900 kN. After the first stage was discarded, the second stage burnt for approximately 6 minutes taking the vehicle and payload to an altitude of 185 kilometres. The second stage was then also discarded.

- Third Stage (S-IVB) weighed 119 metric tons and contained only one J-2 engine which burnt for 2.75 minutes boosting the spacecraft to orbital velocity of about 28,165 km/hour. The third stage was shut down with fuel remaining and remains attached to the spacecraft in Earth orbit. The J-2 engine was later reignited to propel the spacecraft into translunar trajectory (speed of 39,428.928 km/hour) before finally being discarded.

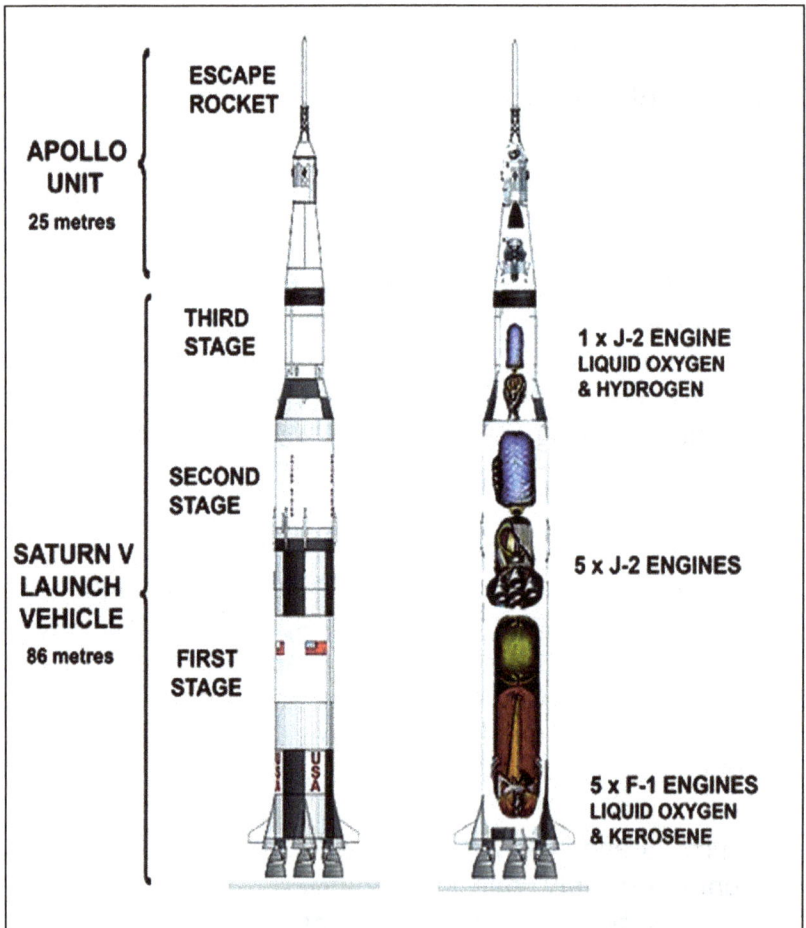

Figure 2.30: The Saturn V Apollo launch vehicle

The Apollo spacecraft consisted of the Command Module (CM), Service Module (SM) and Lunar Module (LM) and the Instrument Unit (IU) which all sat atop the launch vehicle. Above the CM was the Launch Escape System (LES). The Instrument Unit controlled the operations of the rocket from just before lift off until the last stage was discarded. It included guidance and telemetry systems for the rocket. By measuring the acceleration and vehicle attitude, it could calculate the position and velocity of the rocket and correct for any deviations.

On January 27 1967, *Apollo* 1 suffered a disastrous fire on board which killed all three Astronauts (**Virgil "Gus" Grissom** 1926-1927, **Edward H. White** 1930-1967, and **Roger B. Chaffee** 1935-1967). This was probably due to an electrical short within the oxygen-rich atmosphere of the cabin. After re-testing and design, *Apollo* 7 was launched for a multi-orbital flight around the Earth, followed by *Apollo* 8 which flew to the Moon and made 10 orbits of the Moon and returned to Earth. The crews were the first humans to see the far side of the Moon and Earth rise over the lunar horizon with their own eyes.

Figure 2.31: Earth rise from *Apollo* 8 taken on Dec. 24, 1968 (Photo:NASA)

*Apollos* 9 and 10 carried out tests on assembling the Command/Service Module and Lunar Module with a dress-rehearsal around the Moon by *Apollo* 10.

Launched by a from Kennedy Space Centre at Merritt Island, Florida, on July 16, 1969, *Apollo* 11 made one and a half complete orbits of the Earth, taking two hours and forty minutes, before the third stage of the Saturn V rocket ignited pushed the spacecraft into its long trajectory towards the Moon. About thirty minutes later the third stage of Saturn V separated from the *Columbia* (Command/Service Module) and *Eagle* (Lunar Module), the two craft then docked, joined and continued towards the Moon reaching a speed of over 40,000km/hour. The journey towards lunar orbit took three days during which time the astronauts spent working, eating, sleeping, and observing from the Command Module.

On July 19th, *Apollo* 11 passed behind the Moon and fired its service propulsion engine to enter an orbit around the Moon. During the thirty orbits that followed, the crew observed their future landing site in the southern Sea Of Tranquility (Mare Tranquillitatis) about 19 km southwest of the crater Sabine D. The landing site was selected because it had observed as being relatively flat and smooth by the previous robotic landers *Ranger* 8 and *Surveyor 5*.

When *Apollo* 11 achieved the final orbit around the Moon, **Neil Armstrong** (1930 - 2012) and **Edwin "Buzz" Aldrin** (1930 - ) climbed aboard the *Eagle*, undocked from the *Columbia* and began their descent towards the surface; the third crew member, **Michael Collins** (1930 - ) stayed aboard the Columbia. During this time, the automatic landing computer on the *Eagle* malfunctioned and the landing site was seen to be on the edge of a small, rough depression. Armstrong was then forced to manually steer the craft away from its predetermined landing site. In addition, the fuel supply had dropped to a critical level but the craft was able to land safely with just 25 seconds of fuel left. The message *"The Eagle has landed"* was a great relief to Mission Control and the millions of people listening in back on Earth.

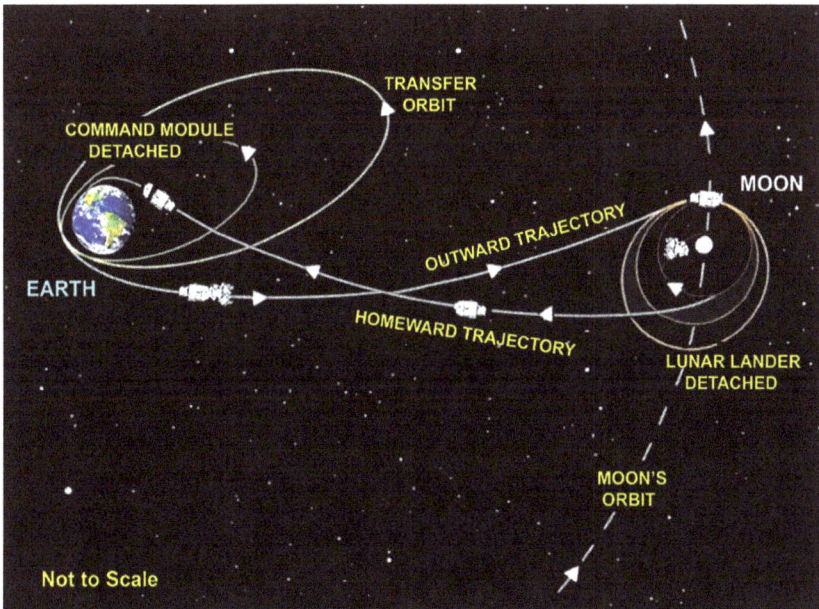

Figure 2.32: Diagram showing the trajectory (pathway) of *Apollo* 11

Armstrong and Aldrin then spent about two and a half hours preparing to leave *Eagle* but, at 02:56:15 **UTC** on July 21, 1969, Neil Armstrong finally set his left foot on the surface. The signal was received at the Honeysuckle Creek Tracking Station in Australia but then minutes later the feed was switched to the more sensitive Parkes radio telescope in

Australia and the first black and white images of the first lunar EVA (Extra Vehicular Activity) were received and broadcast to at least 600 million people on Earth.

Online Video 2.2: First steps on the Moon (NASA)
Go to https://www.youtube.com/watch?v=uJrWFpdnCvI )

"Buzz" Aldrin soon followed and the two Astronauts tested out the Moon's lighter gravity and then proceeded onto their exploration and to set up few small experiments, such as planting a seismograph, planting an American flag and taking photographs. The crew returned to the *Eagle* and, after resting, they launched the upper part of the Eagle from its stable base and re-joined Collins and the *Columbia* in lunar orbit and began their long journey home.

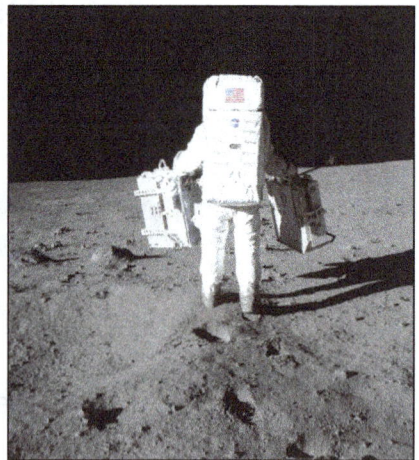

Figures 2.33: & 2.34: Aldrin in front of the *Eagle* – note the gold foil for insulation on the base of the *Eagle* (left) and carrying instrument packages (Photo: NASA)

After the three-day return trip, the astronauts climbed into the small Service Module, jettisoned the Command Module, and re-entered the atmosphere, splashing down in the Pacific Ocean where they were picked up by the aircraft carrier USS *Hornet* at 11:45 a.m. in the Pacific Ocean.

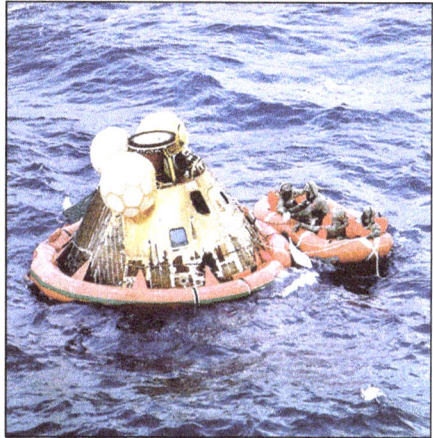

Figures 2.35 & 2.36: Lunar Module *Eagle* flying across the Moon (left) and splashdown in the Pacific Ocean – note the protective suits to prevent anything from the Moon contaminating Earth (Photo: NASA)

Figure 2.37: The *Apollo* 11 Astronauts (from left): Neil Armstrong (1930 – 2012); Michael Collins (1930- ); and Edwin "Buzz" Aldrin Jr. (1930 - ) (Photo: NASA)

There were to be six more Apollo missions to the Moon. Apart from *Apollo* 13 which was aborted due to an oxygen tank explosion in the Service Module, all missions successfully carried out a number of scientific activities. These included:

exploring more areas of the moon
photographing the surface
seismic monitoring
solar wind experiments
geodometer reflectors for Moon-Earth distance calculations
collecting rock samples and drill cores
using an electric lunar rover to move across longer distances

*Apollo* 17 was the last mission to the Moon with the two Astronauts, Eugene Cernan and Harrison Schmitt, leaving the surface on December 14, 1972.

Figures 2.38 & 2.39: At left, Apollo 12's Alan Bean sets up an instrument package on the Oceans of Storms (Photo: NASA/Peter Conrad) and at right is the Radioisotope Thermoelectric Generator which uses heat from a radioactive source to generate electricity for the instrument packages (Photo NASA/Alan Shepard)

Figure 2.40: *Apollo* 15 *Falcon* and its Lunar Rover near the Hadley Rille (Photo: NASA/David Scott).

Figure 2.41: *Apollo* 17's Harrison Schmitt, a Geologist, examines a Large, split rock in the Taurus-Littrow valley (Photo: NASA/Eugene Cernan).

Since leaving the Moon, the American, Russian and European space agencies have focused their attention, often as collaborative missions in building manned space stations and sending robotic probes to other planets, comets and meteors of the Solar System. The Mir, Skylab, Space Shuttle and International Space Station (ongoing) have been the latest of humankind's attempts to colonize space and the Solar System.

Figure 2.42: Skylab orbited Earth from 1973 to 1979 and was allowed to burn up in the Earth's atmosphere over Western Australia (Photo: NASA).

Figure 2.43: One of the several space shuttles re-useable transport vessel and experiment platform (Photo: NASA)

Figure 2.44: The Russian *Mir* ("Peace") space station orbited the Earth from
1986 to 2001 (Photo: NASA)

Figure 2.45: *The International Space Station* has been in orbit since 1998. Note the *Soyuz* re-supply craft (bottom right) (Photo: NASA)

There are many free Apps. available Online (Search Words: NASA, Rockets, Spacecraft – search for Offline Apps. Also search for free eBooks by Jules Verne and H.G. Wells)

# Chapter 3: A Matter Of Perspective

## 3.1 Different Viewpoints

The Sun and its planets have long been admired, worshiped and studied. There had been many theories as to how they all worked. From Earth, the Sun, stars and many of the planets all appeared to rise in the east, travel across the sky along an ordered pathway and then set in the west. Until the Renaissance in Europe, after the 15<sup>th</sup> Century, most educated people thought that the Earth was the centre of the universe and that everything rotated around it. This was the **geocentric theory.** The Greek philosopher **Aristarchus of Samos** (Greek: 310 – c. 230 BC) had speculated that the Sun may be the centre around which the Earth orbited, advocating a **heliocentric theory**, but his ideas were rejected in favour of the more popular Geocentric Theories of **Aristotle** (Greek: 384 – 322 BC) and **Claudius Ptolemy** (Greco-Egyptian: AD 100 – c. 170) which lasted for another 1500 years.

## 3.2 New Discoveries and Ideas

In Europe, Nicholas Copernicus (or Mikolaj Koperni, Polish: 1473-1543) was the first to develop a **heliocentric model** which could be used mathematically to make
predictions about the positions of the planets around the Sun. His famous book, *De revolutionibus orbium coelestium* (*On the Revolutions of the Heavenly Spheres*), first printed in 1543 in Nuremberg (now in Germany), offered an

alternative model of the universe to the geocentric theory which had been widely accepted since ancient times.

Figure 3.1: Nicolaus Copernicus

Later, in the 17th century, **Tycho Brahe** (or Tyge Ottesen Brahe, Danish: 1546-1601), the eccentric but gifted astronomer, had constructed huge devices on the towers of his castle of Uraniborg on the island of Hven to measure and plot the position and angles of planets and stars.

Figure 3.2: Tycho Brahe

From the great wealth of data obtained, he proposed a geocentric theory of the Solar System by combining the mathematical predictability of Copernicus, but still retaining the general notions of Ptolemy. For example, he could make predictions about the planet Mars, but explained its strange orbit which seems to make large loops in the sky as it generally moves westward over several months, in terms of Ptolemy's idea of **epicycles** of small circles moving around a larger circle. Brahe was an interesting character. He was quick tempered and lost part of his nose in a sword duel with a fellow Dane over a maths problem whilst studying in Germany and so wore a brass prosthetic nose. His death at the age of 54 was also unusual in that he died of a bladder infection after attending a banquet in Prague, and had refused to leave the banquet to relieve himself because it would have been a breach of etiquette. Nevertheless, his systematic measurement and lifelong study of the motion of the stars and planets using large, purpose-built instruments earned him a place as one of the great astronomers of history.

**Johannes Kepler** (German: 1571 – 1630) worked in Prague, now in the Czech Republic, as Brahe's long-suffering assistant for the last few years of the great man's life and was able to use the large volume of data which had been gathered by Brahe to calculate how the heavenly bodies actually moved. Unlike Brahe, Kepler saw merit in the heliocentric theory of Copernicus, and fighting against approaching blindness, he formulated his great **laws of planetary motion** which finally described how the planets moved around the Sun, publishing his ideas between 1609 and 1619. These laws state that:

Figure 3.3 Johannes Kepler

**The orbit of a planet around the Sun is an ellipse with the Sun at one of the two foci.** This differed from Copernicus' idea that the orbits of the planets were circle. It is surprising that Kepler, being true to his scientific calculations, and describing the orbits as elliptical, spent most of the rest of his life attempting to fit his calculations within Plato's Greek philosophy that the planets orbited in accordance with the regular solids of Greek geometry (such as tetrahedrons, spheres, cubes).

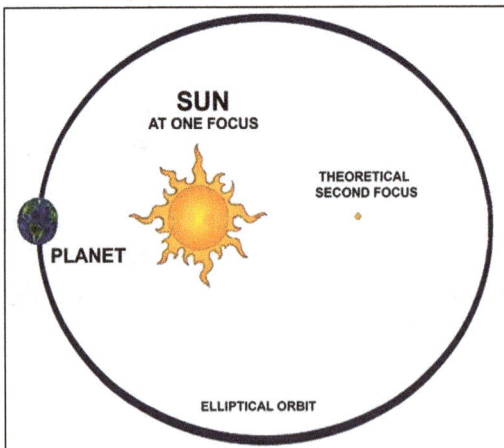

Figure 3.4: Kepler's First Law of Planetary Motion

A line segment joining a planet and the Sun sweeps out equal areas during equal intervals of time. This means that to cover the equal area at perihelion (closest to the Sun), a planet will have to travel faster to cover the longer distance of that part of the orbit in the same time as when it is at aphelion (furthest from the Sun) and only has to cover a relatively short distance in the same time. This is another divergence from Copernicus who believed that a planet travelled at a uniform speed around the Sun.

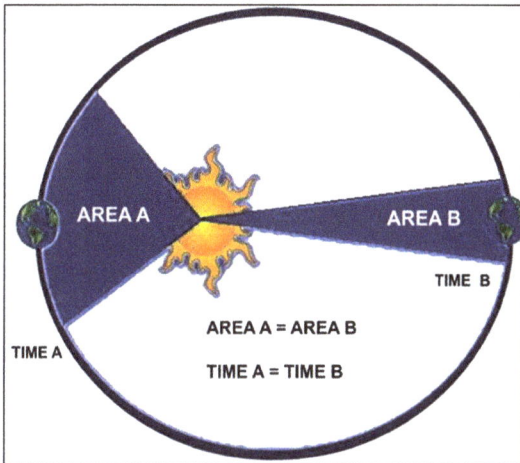

Figure 3.5: Kepler's Second Law of Planetary Motion

The square of the orbital period of a planet is proportional to the cube of the semi-major axis of its orbit (i.e. $T^2 \propto R^3$) This means that Earth-bound astronomers could time the orbital **period** (the time it takes for one complete orbit) of a planet from the time it is in a particular position (or co-ordinate) until it returns to the exact place (this may take years!), and then calculate the planet's distance to the Sun.

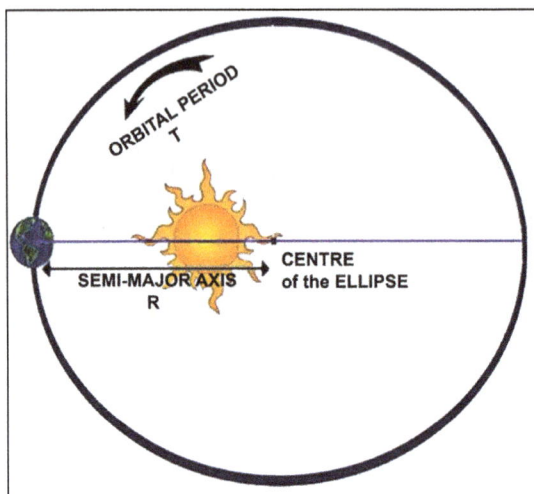

Figure 3.6: Kepler's Third Law of Planetary Motion

In 1604, Kepler also suggested that the spreading of light from a point source obeys an inverse square relationship. This **inverse square law** is now one of the cornerstones of calculations about the radiating of light (and all other types of radiation):

**Intensity (I) is proportional to the inverse of the square of the distance (d) from the source**

or

$$I \; \alpha \;\; 1/d^2$$
($\alpha$ means "is proportional to")

For example, if an observer moves twice the distance from a light source (d=2) then the intensity or brightness of the light ($=1/2^2$ = ¼) will be quartered.

Early in 1610, **Galileo Galilei** (Italian: 1564-1642) used his newly-constructed telescope to observe the bright planet Jupiter and discovered that it had four satellites orbiting it. In the previous year he had constructed this simple telescope based on some uncertain descriptions of the first practical telescope developed in 1608 by **Hans Lippershey** (Dutch: 1570-1619). Galileo's first telescope had only about 3x magnification, but he later made an improved version with a 30x magnification. His observations of the four larger moons of Jupiter, now called the Galilean Moons, orbiting their parent planet, was a startling revelation at a time when most authorities followed the well-established and traditional teachings of **Aristotle** (Greek: 384 – 322 BC) that everything orbited the Earth in its perfection. In 1610 he published his observations in a paper called *Sidereus Nuncius* (*Starry Messenger*) and sought the opinion of Kepler. Kepler responded, endorsing Galileo's observations and suggested several points of

Figure 3.7: Galileo Galilei

Figure 3.8: Galileo's telescopes

view about his observations and the use of the telescope. Later that year, Kepler used his own telescope, which used two double convex lenses instead of Galileo's concave-convex system, to make observations of the Moon which provided further support of Galileo. Unfortunately, in Italy, Galileo's practical evidence for Copernicus' heliocentric theory met great disapproval from elements of the Roman Catholic Church who followed the Aristotelian approach and found Galileo's views heretical and a threat to the existing world order. Consequently, Galileo's works were banned; he was tried and put under house arrest. One of Galileo's followers, a Dominican friar, **Giordano Bruno** (Italian: 1548-1600) was burnt at the stake in 1600 for holding the similar heretical views.

In England, **Sir Isaac Newton** (English: 1642 – 1727) physicist, mathematician and astronomer published his many findings and calculations on motion in his celebrated book, *Principia*. In it, he outlined his famous three **laws of motion** which stated that:

Figure 3.9 Isaac Newton

**First Law: Every object in a state of uniform motion tends to remain in that state of motion unless an external force is applied to it.** In other words, a stationary object will not move nor will a moving object change its speed or direction unless an external force (or push) is applied to it.

Thus in space, a rocket can accelerate to a specific velocity (speed in a given direction), turn off its motors and then continue at that velocity until another force is applied.

**Second Law: The acceleration of an object as produced by a net force is directly proportional to the magnitude of the net force, and acts in the same direction as the net force, and is inversely proportional to the mass of the object.** This means that this acceleration (an increase in velocity per time) increases by predictable amounts if the force is increased but that this acceleration also depends on the mass such that a large mass will accelerate more slowly than a smaller mass if the same force is applied. This can be expressed mathematically as.

$$\text{Acceleration (a)} \; \alpha \;\; \text{Force (F)/ Mass (m)}$$

As it is the force that is more often calculated, this relationship is usually written as:

$$F \;\; \alpha \;\; m \, a$$

That is, the force is directly proportional to the product of the mass and the acceleration and if the appropriate units are used the proportionality relationship disappears and the relationship becomes an equation expressed as:

$$F \;=\; m \, a$$

Where **F** is in **newtons** (N - i.e. 1 kg m/s$^2$);
 **m** is in **kilograms** (kg); and
 **a** is in metres/second/second (m/s$^2$)

**Third Law: To every action there is an equal and opposite reaction.** This usually concerns the **momentum** of equal and opposite forces. For example a rocket with a certain large mass will have a much smaller velocity than the total mass of the exhaust gases which are escaping at high velocity from its motor.

Newton often used this term momentum which is defined as that possessed by a mass with a velocity:

**Momentum (P) = Mass (m) x Velocity (v)**

and as acceleration is the change in velocity per time, Newton's Second Law can be written as:

$$F = P/t$$

where **F** = Force (N)
**P** = Momentum (in kg m/sec) and
**t** = time (seconds)

Newton's most famous contribution to astronomy however, was his work on the concept of **gravitation**. He reasoned that all objects on Earth fall down towards its centre, and that this force of gravity was one of attraction which depended upon the masses of the objects involved (say, the Earth and an apple), but inversely upon the square of the distance between them. He derived this inverse-square relationship because (like light radiating out from a point) gravity would concern distance from the centre of the Earth (a sphere) and masses on its surface:

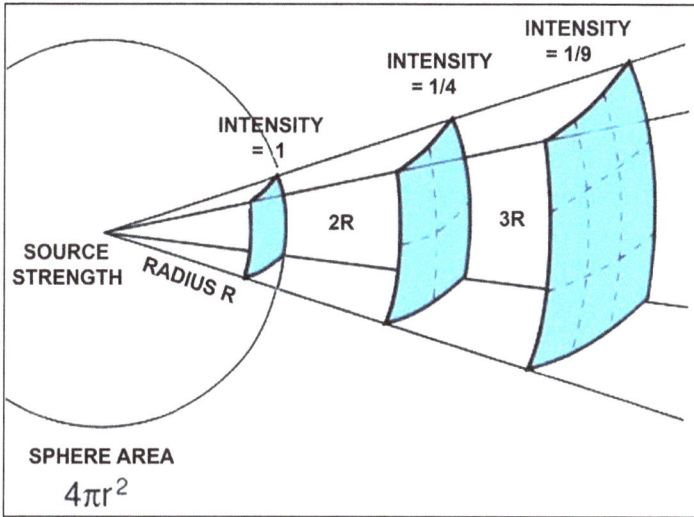

Figure 3.10: The inverse square relationship of gravity

This can be expressed mathematically as:

$$F \; \alpha \; \frac{M1 \; M2}{r^2}$$

where  **F**  is the  Force of Gravity
**M1** and **M2** are the masses involved and
r is the distance between the masses (the
Earth's radius if referring to
gravitational attraction on its surface)

It was not until 1798, over 100 years later, that **Henry Cavendish** (English: 1731-1810) performed an elaborate series of experiments using large metal spheres held near each other and at different distances using a very delicate torsion balance to measure their force of attraction, that

Newton's **law of universal gravitation** was able to be expressed as a mathematical equation:

$$F = G\frac{M1\ M2}{r^2}$$

Where  **F** is the Force in newtons (N)
**G** is the Gravitational Constant
$(6.674\times10^{-11}\ N\,m^2 / kg^2)$
**M1** and **M2** are masses (kg) and
**r** is the distance between the bodies (metres).

Now the radius of the Earth had been known since **Eratosthenes of Cyrene** (276 - 194 BC) had used angles of the Sun's shadow at two different places (although this had been more accurately measured later) and that the acceleration due to gravity (g) for any falling mass (it is a constant- thanks to Galileo proving it experimentally) can be measured directly and is found to be 9.8 m/s$^2$. So for a dropped mass (m):

$$F = mg = \frac{GmM}{r^2}$$

so cancelling out m on both sides of the equation and changing it to give M as the main subject, the mass of the Earth (**M**) could then be calculated:

$$g = \frac{GM}{r^2}$$

therefore   $M = \frac{r^2\,g}{G}$

Now this idea could be extended to the Sun and other planets using some very elegant mathematics and a known equation for the force keeping objects spinning in a circle - **centrifugal force** - like a mass swung in a circle on the end of a rope - such that this force = mass x square of the velocity divided by the radius of the orbit, could be applied to planetary motion. It was reasoned that planets are kept in their regular orbits by the <u>inward</u> attraction of the gravitational pull of the Sun but kept from doing so by the <u>outward</u> centrifugal reaction force due to their speeds going around the Sun given to the planets at the very formation of the spinning disk which formed the Solar System:

$$\text{Centrifugal Force} = \text{Gravitational Force}$$

$$\text{or} \quad F = \frac{mv^2}{r} = \frac{G\,mM}{r^2}$$

where  **m** = mass of the planet
**M** = mass of the Sun
**r** = distance between them and
**v** = orbital velocity of the planet.

Cancelling out m and an r from both sides of the equation gives the relationship:

$$v^2 = \frac{G\,M}{r}$$

Now the orbital velocity of the Earth **(v)** could be found from the length of its orbit (i.e. the circumference of the orbit – assuming it as a circle ) and the time that it takes to make one complete orbit ( a year) and remembering that velocity is distance divided by time or:

$$v = \frac{2\pi r}{T}$$

where $2\pi r$ = the circumference of a circle and
$T$ = the time or period for one orbit (year)

So, knowing the orbital velocity of the Earth, its distance to the Sun and the value of G (all known by the 17th Century from simple geometry and time calculations), the mass of the Sun could be calculated. Now this method could also be used to calculate the mass of any other body in the Solar System.

These laws of motion and universal gravitation, dominated scientists' view of the physical universe well into the 20th Century and using these principles to account for the trajectories of planets, comets and other phenomena, Newton firmly established validity of the heliocentric model of the Solar System, opening up new research areas for other great astronomers. The invention of the telescope led to the discovery of further planets, moons, asteroids and comets and the use of unmanned spacecraft enabled the closer investigation of geological features and conditions on the surfaces of these Solar System bodies.

# Chapter 4: More on Telescopes

## 4.1 Newton's Optics

As well as doing much of the hard work in explaining why the planets went around the Sun in regular orbits, Sir Isaac Newton also experimented in optics, the study of light. He reasoned that a concave mirror which is curved inwards like those used for magnification for shaving, would be better than a convex lens which has sides which are

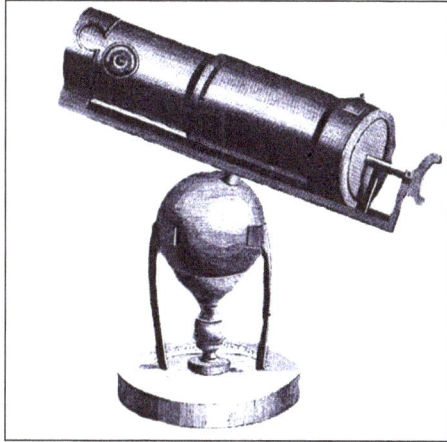

Figure 4.1: Isaac Newton's reflecting telescope

curved outwards and also used for magnification by Kepler in his refracting telescope, itself an improvement of Galileo's simple telescope. Accordingly, in 1672 he presented his new reflecting telescope, later called a Newtonian telescope, to the Royal Society. It had a six inch (15 cm) concave primary mirror made of a polished metal called speculum, an alloy of tin and copper, which could be shaped well to give a highly polished, reflecting surface. The telescope gave a magnification of about forty times

In 1666, in a dark, dusty room with a narrow beam of sunlight entering through the curtains, Newton observed that a triangular glass prism gave a **spectrum** of colours as the beam of light was refracted, or bent, through it. Each colour seemed to be refracted at a different angle. This led him to conclude that colour is a property intrinsic to light and that colour is the result of objects interacting with already-coloured light rather than non-luminous objects generating the colour themselves. This is known as **Newton's theory of colour.** The general effect of splitting sunlight (called white light) into its component colours of different wavelengths is now called dispersion, and the colours are identified as red, orange, yellow, green, blue, indigo and violet – the visible spectrum.

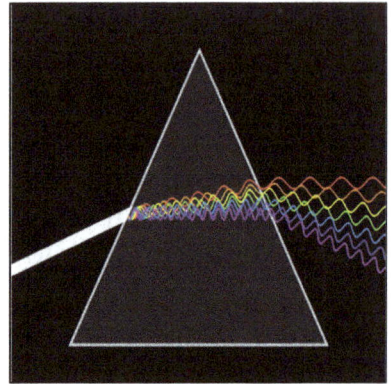

Figures 4.2 & 4.3: A triangular glass prism dispersing sunlight (left) and a diagram showing the refraction and dispersion of light (right)

During the early 1800s, **Joseph von Fraunhofer** (German: 1787 – 1826), using Newton's experiment of the dispersion of light by a prism, developed the analytical method of **spectroscopy** by which light from a luminous source such as a star or coloured flame, could be broken up into its characteristic spectrum. **Robert Bunsen** (German: 1811-1899) and **Gustav Kirchhoff** (German: 1824-1887) used the spectroscope to examine the light emitted from chemical elements burning in a gas flame. They discovered that each element had a characteristic spectrum (as bands of separated colours) which could be used for its identification and in 1868 **Jules Janssen** (French: 1824-1907) discovered the element helium (not known on Earth at that time) from new spectral lines from the Sun. Since then, spectroscopy has played and continues to play a significant role in astronomy as well as in chemistry and physics. Attached to the eyepiece position of a telescope, a **spectroscope** will give a pattern of lines, a **spectrogram,** as the characteristic spectrum for a star or any luminous object in the sky and provide detail of its chemical composition.

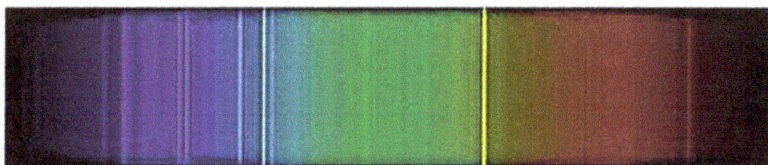

Figure 4.4: An emission spectrum (bright lines) for the element Helium superimposed on the visible spectrum from the Sun. Another type of spectrum is the absorption spectrum which shows dark bands on a full coloured spectrum background instead of these bright bands

## 4.2 Types of Telescopes

With further advances in optics and the concept of electromagnetic radiation other than light, a variety of telescopes were developed from the 17th Century. These included:

Optical Telescopes use lenses and mirrors to form an image. There are several types of optical telescopes:

- **Refracting telescopes** which use ground curved lenses to collect light to form an image and then magnify it. Galileo's telescope used a convex or converging lens as the main light-collecting primary or objective lens. This usually is of long **focal length,** which is the distance between the lens and its focused image, and creates the image of the object in front of the eyepiece. The eyepiece (also a convex lens) is then placed near the position of this image but within its focal length so as to give a magnified image (just like in a magnifying glass). In Galileo's telescope, the eyepiece was a plano-concave lens with one side flat and the other curved inwards. This had the advantage of producing an image which is the correct way up but being only a small lens, had a very narrow field of view.

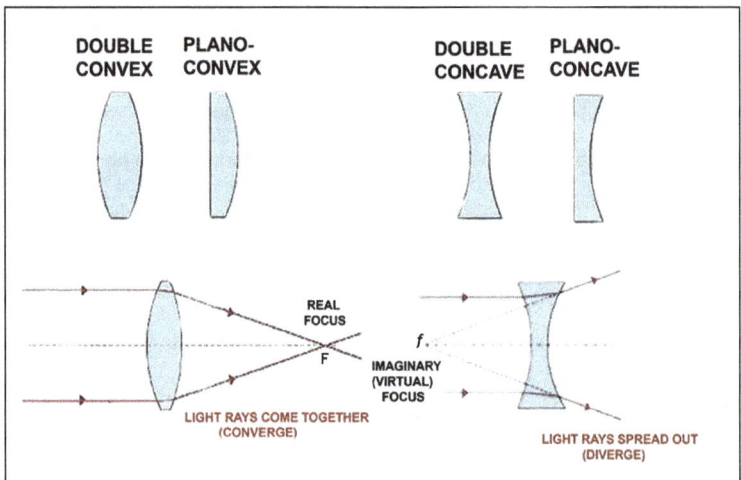

Figure 4.5: Diagram showing simple lenses and their effect on light

The disadvantage of the refractor is that the wider the diameter objective (to gather more light for a brighter image), the longer the focal length (to avoid distortion) and so the longer the telescope. Also, glass lenses can suffer from dispersion and distortion defects.

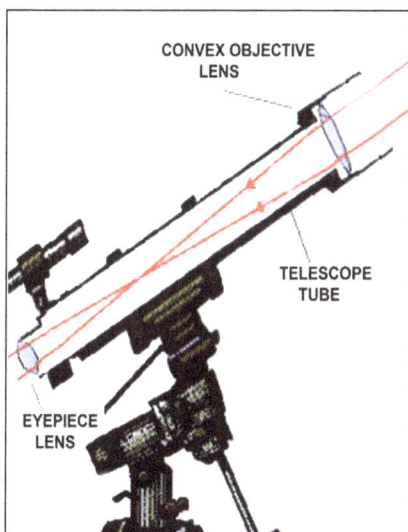

Figures 4.6 & 4.7: A simple refracting telescope (left) and a cut-away view showing its construction (right)

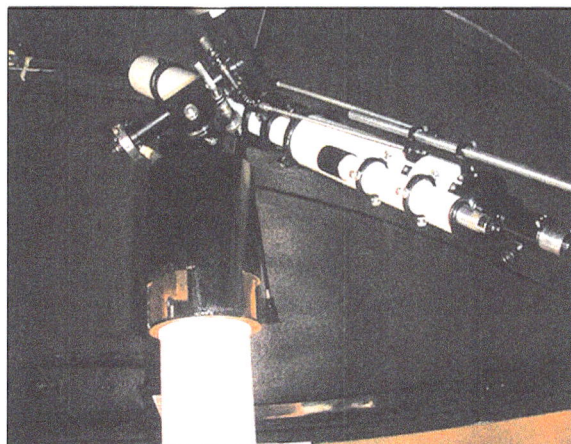

Figure 4.8: A refractor with a 127 mm (5 inch) objective but a length of over 3 metres.

- **Reflecting telescopes** use a curved mirror to gather the light and make an image and a lens eyepiece to magnify this image as in refractors. Newton used a concave mirror made of a highly-polished metal alloy (**speculum** - two-thirds copper and one-third tin) to demonstrate that it would not have the problems of glass lenses like refracting telescopes. Modern reflectors often use parabolic mirrors to give a better image.

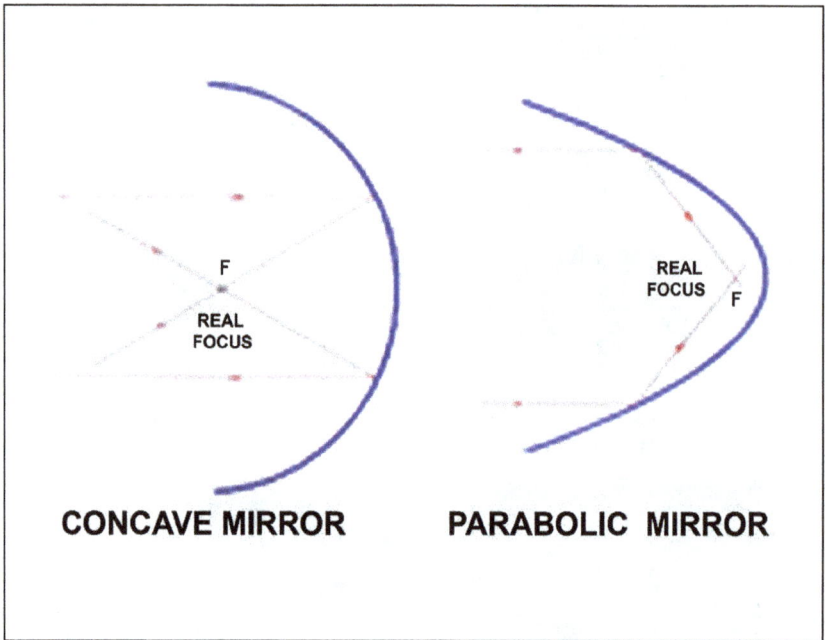

**CONCAVE MIRROR**      **PARABOLIC MIRROR**

Figure 4.9: Diagrams showing the difference between concave and parabolic mirrors

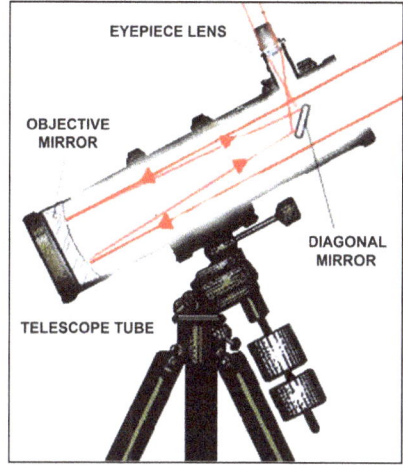

Figures 4.10 & 4.11: Diagrams showing a Newtonian reflecting telescope and a cut-away view

Whilst the Newtonian telescope reduces some of the lens defects of the refractor and allows for the construction of bigger mirrors for better light gathering, it has the disadvantage that the eyepiece is placed up near the front of the opening or **aperture**. In some large reflectors, the observation position is actually up in the front of the telescope near the primary focus of the mirror.

Figure 4.12: A small 203 mm (8 inch) Newtonian reflector. Note the position of the eyepiece.

A more efficient, modified reflecting telescope is the **Cassegrain reflector.** This uses a parabolic primary mirror with a convex secondary mirror to re-direct and focus the light through the rear of the telescope, through the primary mirror to the eyepiece. This design puts the focal point at a convenient position to the rear of the telescope and the
 convex secondary mirror extends the focal length of the primary as well as giving additional magnification. This also allows for a wide aperture which increases the light-gathering, or **resolution**, and clarity of the telescope. This type of telescope was developed by **Laurent Cassegrain** (French: 1629-1623) in 1672, but an improved version, the Schmidt-Cassegrain was later developed for astrophotography in the 1940s and these are often called a **catadioptric telescope** because they use a combination of corrected lenses and mirrors to obtain a very clear image.

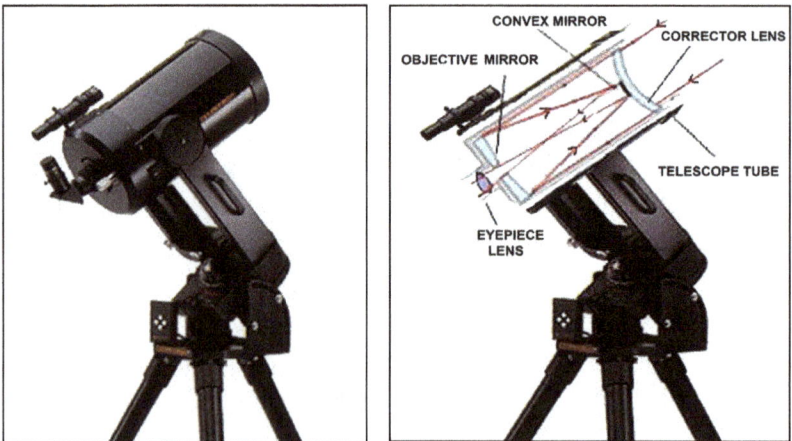

Figures 4.13 & 4.14: Diagrams showing a Schmidt-Cassegrain reflecting telescope (left) and a cut-away view (right).

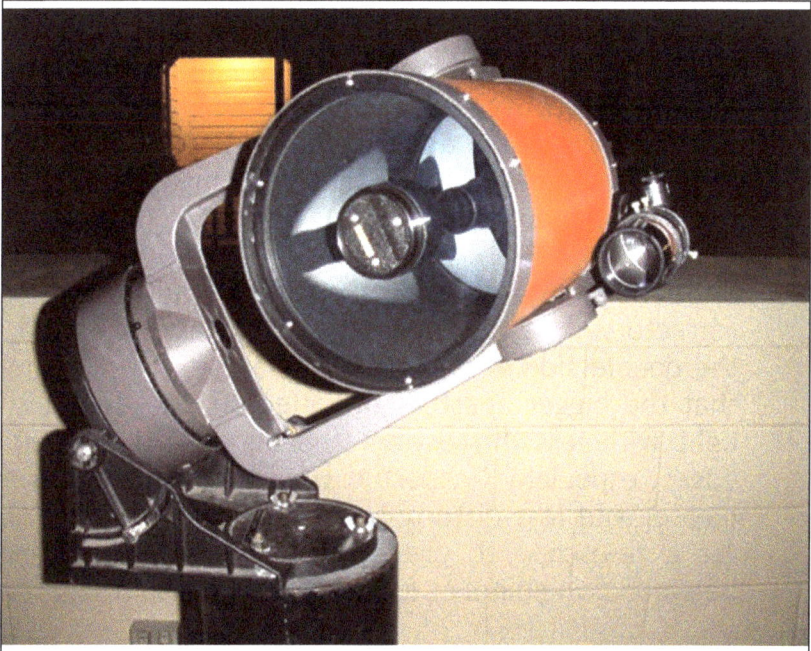

Figure 4.15: Another view of a Schmidt-Cassegrain catadioptric telescope

The **magnification** of any telescope can be found by dividing the focal length of the primary (lens or mirror) by that of each eyepiece.

Magnification = <u>focal length of primary lens or mirror</u>

focal length of the eyepiece

So if a refractor has a focal length of 1000 mm and the eyepiece has a focal length of 10 mm, then the magnification will be 1000/10 = 100 times.

Sometimes a large magnification is a disadvantage the eyepieces will often have a small diameter and so they will pass less light - higher magnification usually means dimmer vision. Sometimes a telescope's optical power will be described in a unit called a **dioptre**. This is equal to the reciprocal of the focal length measured in metres (i.e. 1/focal length in metres).

Reflecting telescopes have many advantages over refractors. A large, curved mirror is easier to grind than the double side of a lens of comparable size. This means that the bigger aperture of reflectors can draw in more light with less **diffraction** (the bending of light around edges, especially in apertures) and so the image is much clearer and less distorted than in refractors. The ability for a telescope to give a clear image and distinguish between two points at a distance is its resolution.

A good guide to the resolving power of a telescope measured in angular units of **seconds of arc** is given by **Dawes limit**:

**R = 116/D (note R = 4.56/D if D is in inches)**

where **R** = resolution in seconds of arc
(60 seconds of arc = 1 degree)
**D** = the diameter of the telescope's objective lens or mirror measured in millimetres (mm).

For example, a 120 mm aperture reflecting telescope will have a resolution of:

**R = 116/120mm**

**= 0.96 seconds of arc**
(or arc sec.)

So, with a resolution of about 1 arc second, this telescope would be able to distinguish between two objects on the Moon about 114 km. apart (since the separation distance between the two points is given by the tangent of the angle of resolution in degrees times the distance to the Moon).

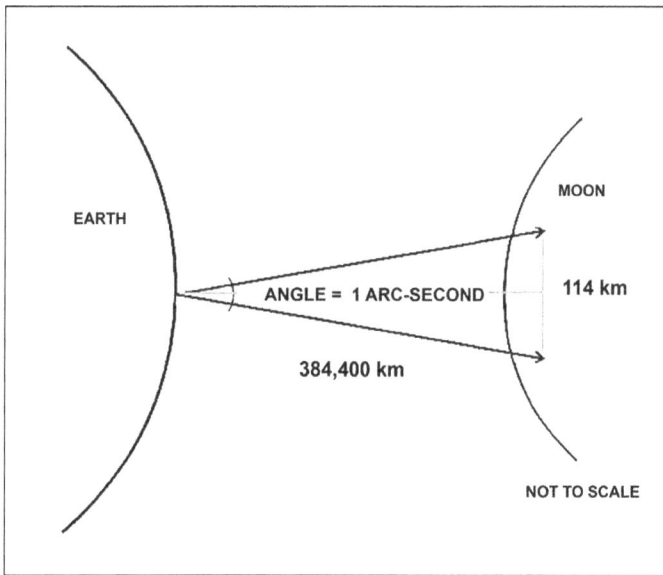

Figure 4.16: Diagram showing the use of angular distance and resolution

In addition to the better resolution of reflectors, they suffer fewer optical defects due to the manufacture of the mirrors and lenses. Both mirrors and lenses can have **spherical aberration**, giving a distorted image, especially around the edges, because of the uneven nature of the ground surface. Lenses also have **chromatic aberration,** because being made from regular, transparent materials, they will also disperse light as well as refract it. This is seen as an image with coloured edges. Most good telescopes have high-quality compound lenses called **achromatic lenses**, which have an additional correcting lens joined with others to reduce the dispersion.

Figure 4.17: Photograph taken through a lens with chromatic aberration

After 1945, there had been some serious proposals to launch a telescope into space so that the distortion of the Earth's atmosphere and light pollution from cities could be avoided. The first ultraviolet spectrum of the Sun was photographed from a low-orbit rocket in 1946, and the National Aeronautics and Space Administration (NASA) launched the *Orbiting Solar Observatory* (OSO) in

1962 to obtain UV, X-ray, and gamma-ray spectra. In that same year, an orbiting solar telescope was launched by the United Kingdom as part of the Ariel space program and four years later, NASA launched the first *Orbiting Astronomical Observatory* (OAO) mission but this failed after three days. The OAO-2 was launched in 1968 and carried out ultraviolet observations of stars and galaxies until 1972.

The *Hubble Space Telescope* (HST) was launched into low Earth orbit (between 160 and 2000 km. altitude) in 1990, and is still sending clear photographs back to Earth via radio links. The telescope is named after the astronomer Edwin Hubble (American: 1889 – 1953). It is a Cassegrain-type telescope with a 2.4-metre primary mirror and images are processed and transmitted back to Earth. It has four main instruments for observing in the near ultraviolet, visible, and near infrared spectra. The HST is one of several space telescopes launched into in Earth orbit in recent times, including NASA's *Compton Gamma Ray Observatory* (1991-2000), the *Chandra X-Ray Observatory* (1999- ), the *Spitzer Space Telescope* (2003 - ) and the European Space Agency's *Herschel Space Observatory* (2009-2013). The replacement for the Hubble Telescope is the *James Webb Space Telescope* (JWST) scheduled for launch in 2018. This telescope features a segmented 6.5-meter diameter primary mirror and will be located near the Earth–Sun **Lagrange point** $L_2$ -the position between two bodies, such as the Earth and Sun, where a third object such as this telescope will not experience any gravitational effects due to the other bodies. It will be further from Earth at a distance of 1.5 million kilometres.

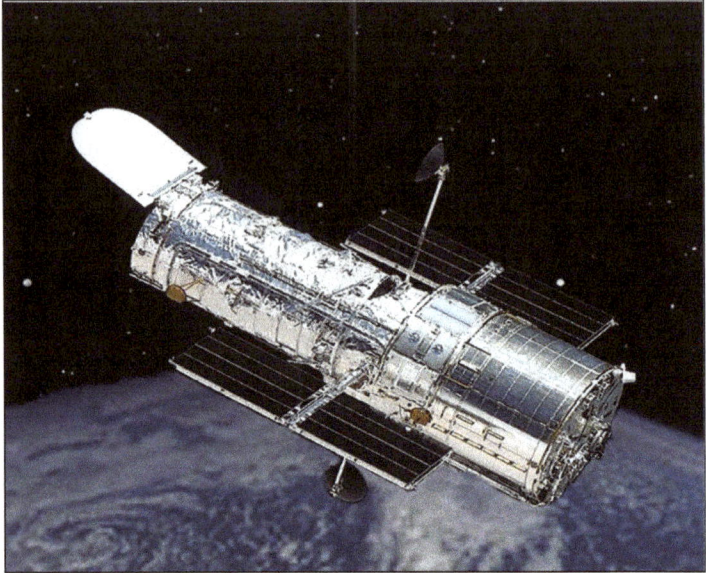

Figure 4.18: The Hubble Space Telescope (Photo of basic craft: NASA)

- **Radio telescopes** began in 1932, when **Karl Jansky** (American: 1905 – 1950), an engineer with the American Bell Telephone Laboratories, was asked to investigate possible sources of interference which could affect the company's new short wave transatlantic radio telephone service which had wavelengths of about 10-20 metres. Subsequently he erected a series of antennae and after several months, recorded a variety of radio signal which he identified as nearby thunderstorms, distant thunderstorms, and a faint steady hiss of unknown origin. This unknown signal was found to be repeated every 23 hours and 56 minutes - the period of a **sidereal day** or one Earth rotational period. Jansky suspected that this background noise originated well beyond the Earth's atmosphere, and concluded that the radiation was coming from our own Milky Way Galaxy and was strongest in the direction of the centre of the galaxy in the constellation of Sagittarius.

Figure 4.19: Karl Jansky and his original antenna (Source: NASA)

In 1937, **Grote Reber** (American: 1911 – 2002) a radio engineer and amateur astronomer, built the first radio telescope, along the lines of Jansky's antenna array, in his backyard and did the first systematic survey of astronomical radio waves. After the Second World War, Professor **John D. Kraus** (American: 1910 –2004) started a radio observatory at Ohio State University and opened up a new tool for deep space observations in astronomy. Since that time, radio astronomy has become the main tool of deep space observation, with many larger, single parabolic dish antennae and multiple-dish arrays being built. Multiple arrays consist of a series of smaller dishes set out in a pattern over a very wide area and linked so that their signals are combined as one powerful input. They can even be linked across the diameter of the Earth or used at different times and positions in the Earth's orbit to act as one gigantic antenna. The signals which are received by the individual units of the array are combined using interferometers which are devices which can combine several radio signals into one. Radio telescopes could now look further out into space than optical telescopes which are restricted to the wavelengths of visible light. The incoming radio signal can be amplified and displayed as sound or a visual signal on a cathode ray oscilloscope screen. Later improvements allowed these signals to be displayed as computer-generated, false-colour images. Interpretation of these images could then be done by comparing them to similar but known visible objects giving a similar radio output.

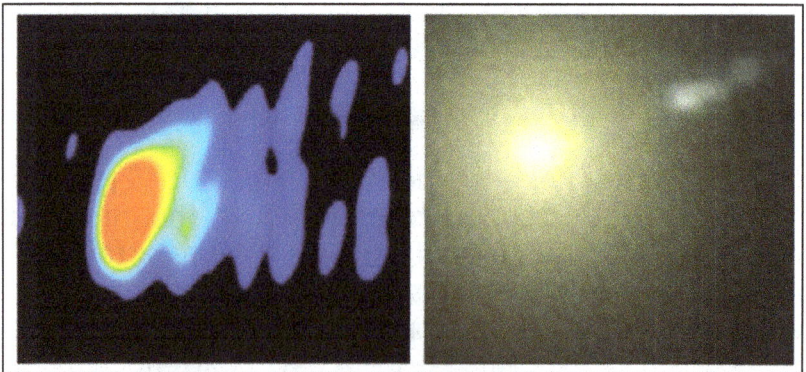

Figure 4.20: Two views of the Galaxy M87 showing the false radio image (left) and its optical image (right) (Photos: NASA)

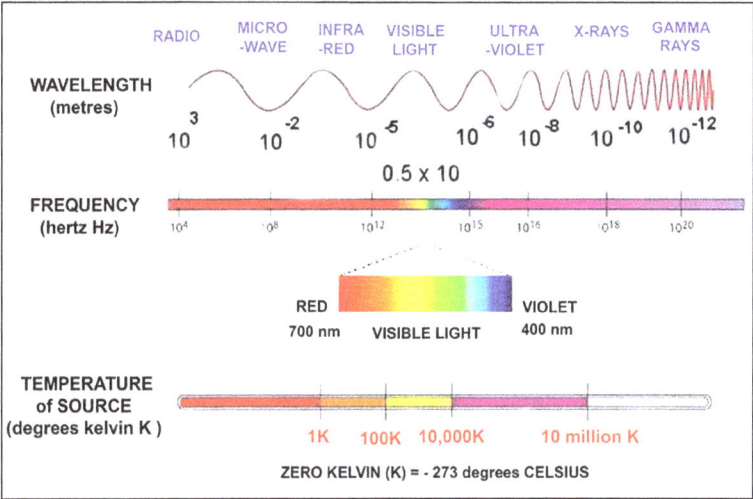

Figure 4.21: Diagram of the electromagnetic spectrum showing the relative positions of radio waves and visible light

Figure 4.22: The 64 metre radio telescope disk near Parkes, New South Wales, Australia

# Chapter 5: More of the Solar System

## 5.1 Perspective and Location

The Solar System consists of the Sun, a medium-sized yellow **star** around which orbit eight major planets, a number of smaller or dwarf planets, asteroids, meteors, comets and much smaller debris left over when the Solar System was formed about 4.5 billion (4,500,000,000) years ago within our **galaxy,** which is a very large group of stars called the Milky Way. Further out from the major planets is the **Kuiper Belt,** named after Dutch-American astronomer Gerard Kuiper: 1905-1973, which contains the minor planets, many asteroids and frozen ices of gases such as methane, ammonia and water. All of these bodies orbit the Sun in slightly elliptical orbits on a flattened disk or plane called the **ecliptic.** Beyond that, extending over the whole Solar System as a sphere at a great distance, about 14,960 billion kilometres from the Sun, marking the outer limit of the Solar System is the **Oort Cloud** named after the astronomer **Jan Oort** (Dutch: 1900-1992).

Figure 5.1: The approximate position of our Sun in the Milky Way galaxy (Photo: NASA of another spiral galaxy similar to our own)

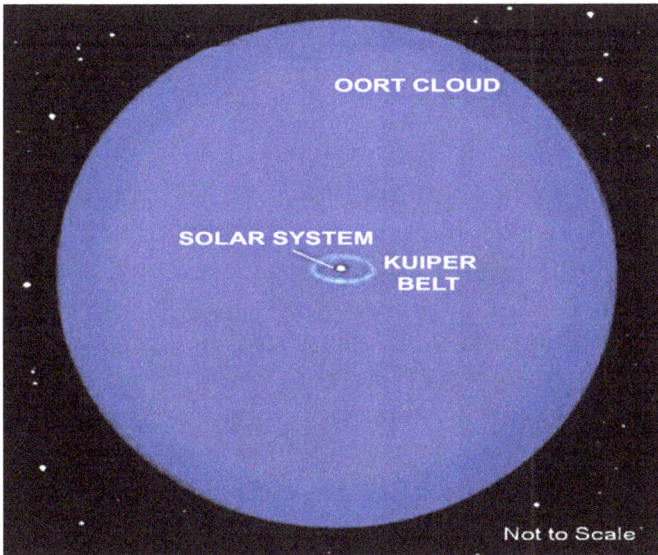

Figure 5.2: An artist's impression of the Oort

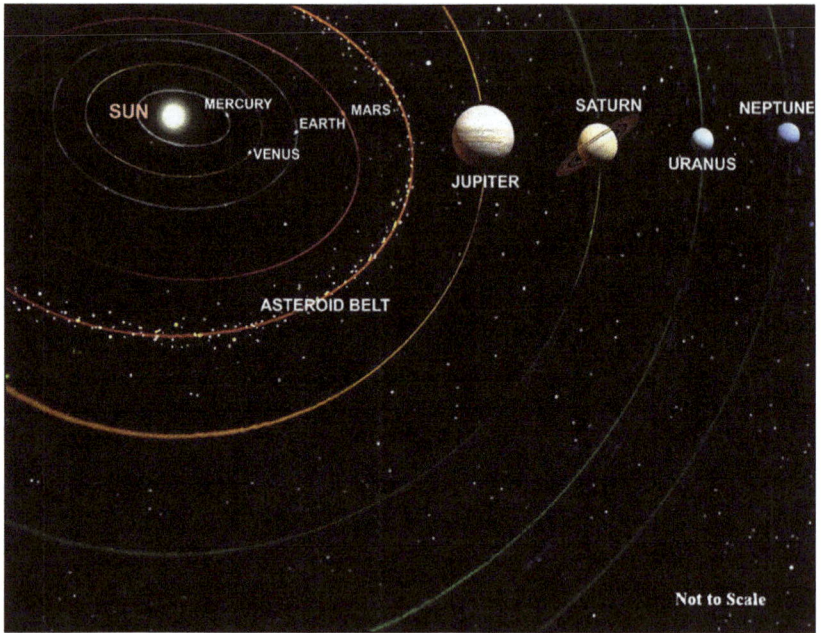

Figure 5.3: Diagram showing the Sun, Asteroid Belt and major planets of the Solar System

The most accepted idea about the formation of the Solar System is the Core Accretion model, an updated version of the earlier nebular hypothesis. This hypothesis was first suggested in the 18th century by **Emanuel Swedenborg** (Swedish:1688-1772), **Emmanuel Kant** (German:1724 – 1804), and **Pierre-Simon Laplace** (French:1749-1827). In this hypothesis, the Solar System formed from the collapse of a giant cloud of interstellar matter within our galaxy, possibly triggered off by the explosion or **supernova** of a giant star nearby. The local gas cloud of mostly hydrogen, helium, dust and ice called the **protoplanetary disk**, began to collapse due to its own gravity which pulled everything in towards its centre of mass. This central core then heated up by gravitational compression to over one million degrees Celsius. At that temperature, thermonuclear fusion began to combine the atomic nuclei of the hydrogen atoms to form helium atoms with the liberation of a great amount of heat, light, nuclear radiation and particles. In time, fusion reactions combined many of the lighter atomic nuclei into larger nuclei, creating heavier atoms. Modern astronomers and astrophysicists suggest that this occurred about 4.6 billion years ago.

Figure 5.4: An artist's impression of the protostar, the Sun, and its disk

The inward gravitational force being slightly off-centre caused the original mass to spin, throwing off spinning concentrations of matter which later formed into the planets, the asteroid belt and other minor bodies. The vast majority of the Solar System's mass, about 99.86% is in the

Sun with much of the remaining mass contained in the largest planet, Jupiter then the rest of the planets, moons and other bodies. The inner planets, Mercury, Venus, Earth and Mars are called the Terrestrial Planets and are composed mainly of rock and metal, whereas the outer planets are composed of lighter materials. Jupiter and Saturn are called the Gas Giants, and are composed of hydrogen and helium as both gas and solid. Uranus and Neptune are the Ice Giants and are composed of a greater amount of the ices of water, ammonia and methane. Combined, the Gas Giants and Ice Giants have more mass than the inner planets.

More recent studies of **exoplanets,** which are planets orbiting other stars outside of our Solar System, suggest that the stable orbits of our planets and other bodies of our Solar System where not always as they are now. Observations of these other solar systems suggest that simple protoplanetary disks often do not last for very long and that large Jupiter-sized planets are often found close to their star. These giant planets are called Hot Jupiters and have very fast orbital periods of only a few days. Some hypotheses suggest that in the early period of our Solar System's formation, perhaps up to the first five billion years, Jupiter would have formed quickly from the same gases found in the Sun as it was much closer to the Sun at that time. As it rapidly moved around the Sun it would have swept away, probably by accretion to itself, most of the inner debris of the early protoplanetary disk. Gravitational interaction with a newly-formed and massive Saturn may have then caused both large planets to move out into their current orbits further out from the Sun. This Jupiter-Saturn interaction, containing much of the mass of the disk, also influenced the nearby Ice Giants, Uranus and Neptune which also were pushed away to the outer regions of the Solar System. Much of the ice, rock and smaller **planetesimals** were either flung out beyond the planets to form the Kuiper Belt and the Oort Cloud beyond, or remained closer to the gravitational pull of the Sun and accreted together to form the inner Terrestrial Planets. Near the end of this chaos, there were many collisions between the new planets and some of the remaining debris left behind. This late heavy bombardment (LHB) of this debris is thought to have caused the craters seen on the Terrestrial Planets and many moons within the Solar System. This **planetary migration hypothesis** would also explain the apparent smaller size of Mars, as Jupiter would have swept away much of the material in its vicinity providing less material for it to form into a larger planet.

## 5.2 The Sun (Symbol: ⊙ A Globe with a Dot for the Centre)

Our Sun (known in the Greek: *Helios*; and in Latin: *Sol*) – a term which directly comes from the Old English *sunne*, probably meaning South. It is a yellow dwarf star about halfway through its life-cycle. It is a nearly perfect spherical ball of **plasma** of hot, electrically-charged material made up of the nuclei of atoms and has internal convection currents in its outer section which generate a strong magnetic field. Its diameter is about 109 times that of Earth and it has a mass about 330,000 times that of Earth, accounting for about 99.86% of the total mass of the Solar System.

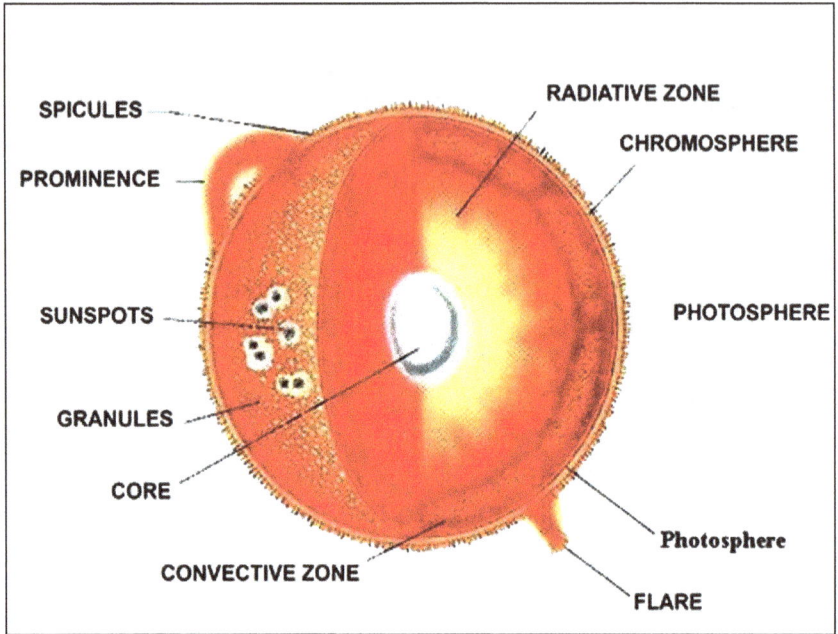

Figure 5.5: The Sun and its interior

The core of the Sun extends from its centre to about 20-25% of the solar radius and is where most of the Sun's energy is generated. From here, the energy is radiated through the **radiative zone,** a place of uniform rotation of the internal plasma, and then across the **tachocline boundary,** to the **convective zone** where heat is transferred by convection currents to the surface, the **photosphere,** causing a granular appearance. Above the photosphere is the first layer of the Sun's atmosphere called the **chromosphere.** This is about 2000 km thick and extends out into the **corona,** the large glow extending far out into space which only be seen during a total eclipse of the Sun. The corona is far hotter than the surface of the Sun (about 3 million degrees Celsius) and this is problematic.

Figure 5.6: A total Solar Eclipse showing the Chromosphere and the outer Corona. Note the plasma streaming out along the magnetic field (Photo: NASA)

Much of the Sun's mass consists of hydrogen (75%) and the rest is mostly helium with much smaller quantities of heavier elements, including oxygen, carbon, neon and iron. The temperature at its centre has been estimated to be about $1.57 \times 10^7$ degrees **kelvin** (K), where $1K = -273°$ Celsius. This heat is generated by the thermonuclear process of **fusion** in which the centres (nuclei) of hydrogen are forced together to become helium nuclei with the expulsion of vast amounts of light, heat, neutrons and other radioactive particles.

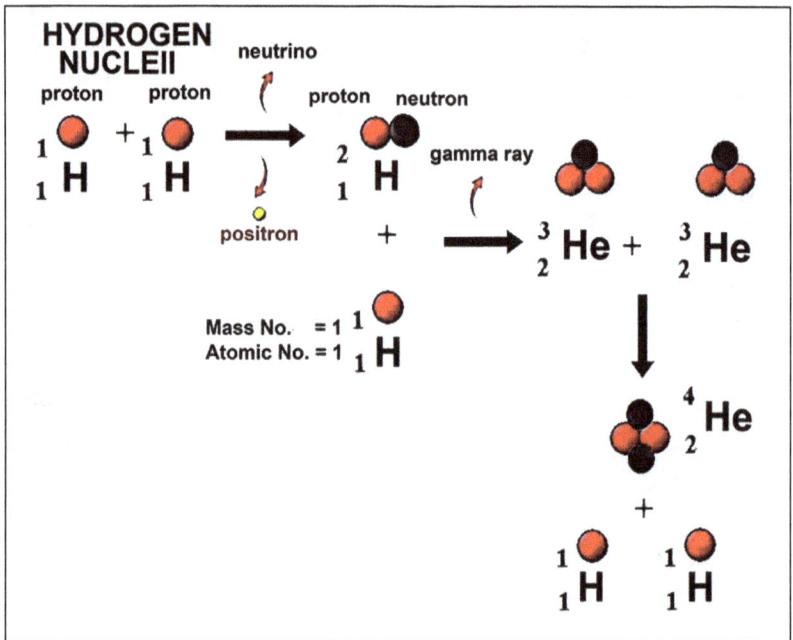

Figure 5.7: A simplified view of nuclear fusion reactions in the Sun.(note: Atomic Number = number of protons; Mass Number = number of protons + neutrons)

In this form of nuclear fusion, hydrogen atom nuclei as protons, combine to form a heavy isotope of hydrogen having an extra neutron with the removal of neutrino and positron particles. This isotope is called deuterium or hydrogen 2. This combines with another proton to form helium 3 **and gamma** rays. Two helium 3 particles combine to form normal helium plus two protons. At each stage there is much heat given off.

The distance of the Earth from the Sun was a major study by early astronomers. Observations and calculations by the Ancient Greeks (notably **Aristarchus of Samos:** 310 – 230 BC) could only give very rough approximations for the distance to the Sun, and then it was often as a ratio to the radius of the Earth. By the 17th Century, distances had been more accurately calculated but they were still expressed as a ratio, and it was convenient to use the distance to the Sun as a unit, later called the **astronomical unit** (**AU**). Whilst Tycho Brahe and Johannes Kepler had attempted to measure the actual distance, it was not until 1663 that the Scottish astronomer and mathematician **James Gregory** (1638-1675) suggested a method for doing this. He reasoned that by measuring the **transit** of a planet, its observed passage across the face of the Sun, such as Mercury from two different places on the Earth, the distance from Earth to the Sun could be calculated. **Edmund Halley** (English: 1656 –1742) used this idea with the transit of Venus of 1769 to make his calculations to solve the same problem. Several observation sites were selected, including sites in Norway, North America and in Tahiti. This latter site was to be undertaken by the astronomer, **Charles Green** (English: 1734-1731) with the aid of the naturalist **Daniel Solander** (Swedish: 1733-1782) during the expedition of **Captain James Cook** (English: 1728-1779) to the Pacific Ocean in 1769.

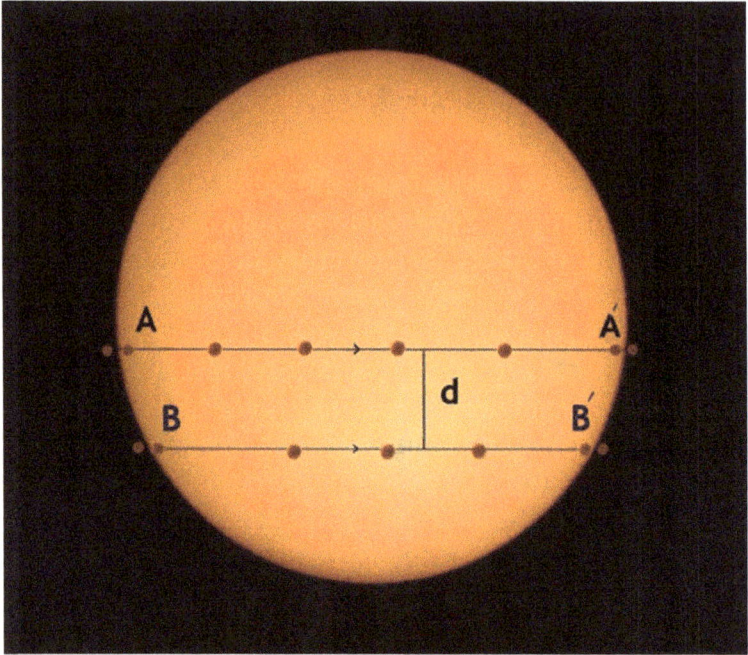

Figure 5.8 Diagram showing the transit of the planet Venus at different times and as seen from two different locations on Earth (giving transits AA' and BB'). "d" is the distance between these two transits on the face of the Sun.

Data from these sites were used with the method of parallax to calculate the angle made from the observation sites on Earth to the transit of Venus across the face of the Sun.

Halley had reasoned that the distance between these two transits (d) would give the angle of solar parallax observed from Earth from at least two sites and that this could be measured using the time taken for both transits. This calculation is very complicated and uses Pythagoras' theorem as well as the period of orbit of Venus and the radius of the Sun. Its use is beyond this text but for interest, a simplified version can be found at:

http://serviastro.am.ub.edu/twiki/bin/view/ServiAstro/Cal culTerrasolapartirDeVenus

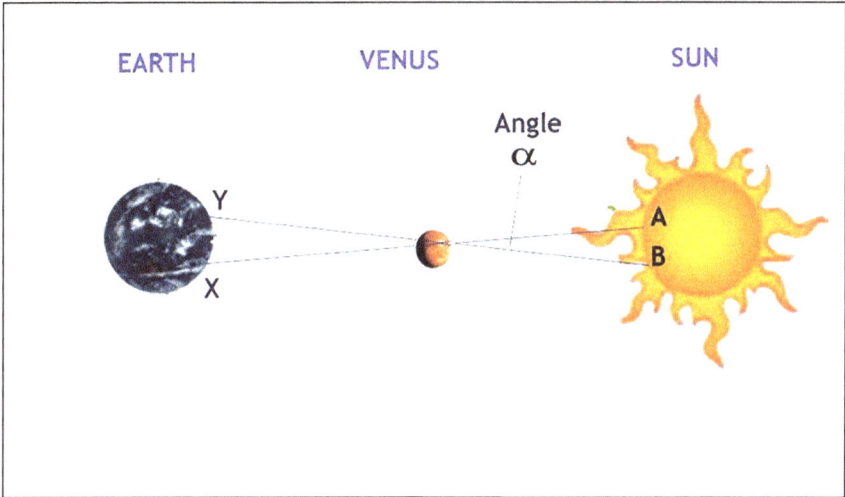

Figure 5.9: Diagram showing the angle of parallax between the Sun Venus and Earth from observation sites X and Y

Having calculated this angle of parallax (α) it is then a matter of trigonometry to calculate the distance between Venus and Earth knowing the distance between X and Y which could be measured on the Earth's surface.

From Figure 5.10, the distance to Venus (OV) can be calculated in known units (e.g. kilometres):

**Tangent of α/2 = OY/OV**

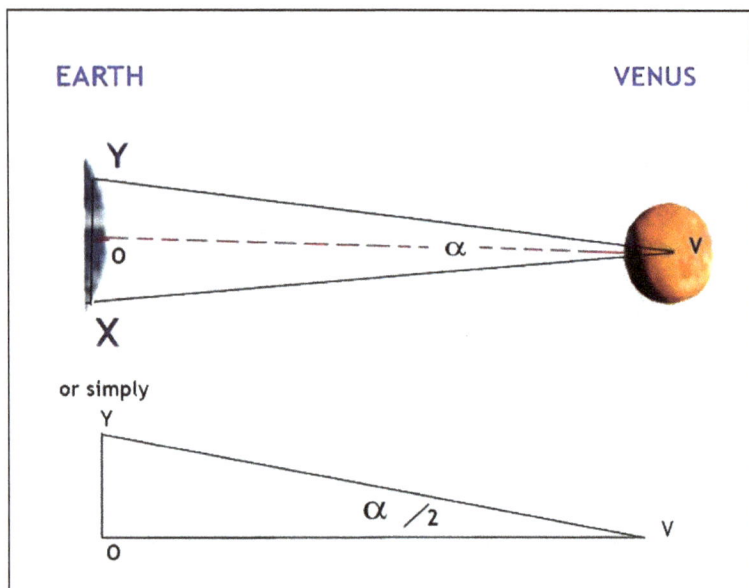

Figure 5.10: Diagram of the trigonometry for finding the distance to Venus

Where:   $\alpha$ has been calculated by the transit and
         OY has been measured on Earth.

i.e. the distance to Venus (OV) = 2OY/$\alpha$

Now, from Kepler's Third Law (i.e. the square of the <u>orbital</u> <u>period</u> of a planet is proportional to the cube of the semi-major axis of its orbit), it had been long known that Venus had a distance to the Sun of 0.72 astronomical units (with

Earth = 1 AU) so it was a simple matter of scaling is value up to get the total distance to the Sun.

The astronomical unit has undergone many refinements since then so that today, the A.U. is taken as the average distance between the Earth and the Sun and has been set by the International Astronomical Union as 149,597,870.700 kilometres (92.955807 million miles).

Some of the main facts about our Sun are:

- average distance from Earth of $1.496 \times 10^8$ km (it takes light 8 min 19 sec to reach Earth from the Sun)

- age estimated at about 4.567 billion years (about the middle age for such a star)

- galactic period (time taken for one orbit of the Milky Way centre) $2.25\text{-}2.50 \times 10^8$ years

- average orbital velocity of 220 km/s

- rotation period (day) 25.38 days

- equatorial rotation velocity of $7.189 \times 10^3$ km/hour

- equatorial radius of $696342 \pm 65$ km (109 × Earth)

- axial tilt (obliquity) of $7.25°$ to the ecliptic

- surface area of $6.09 \times 10^{12}$ km$^2$ (12000 × Earth)

- volume of $1.41 \times 10^{18}$ km$^3$ (1300000 × Earth)
- mass of $1.98855 \times 10^{30}$ kg (333000 × Earth)

- average density of 1.408 g/cm (0.255 × Earth)

- gravitational acceleration of 274.0 m/s$^2$ (28 × Earth)

- escape velocity 617.7 km/s (55 × Earth)

- Magnetic field varies across the surface of the Sun and in places from time to time. Its polar field is 0.0001–0.0002 T, whereas the field is typically 0.3 T within surface disturbances called **sunspots** and 0.001–0.01 T in eruptive **solar prominences** in which masses are thrown out from the surface in great magnetic loops. **Solar flares** are smaller eruptions from the surface. There seems to be an 11-year cycle where sunspots and magnetic field disturbance are at their maximum number and size. When this occurs there is usually a corresponding disturbance in the Earth's ionosphere with bigger auroras and disturbances to radio transmissions.

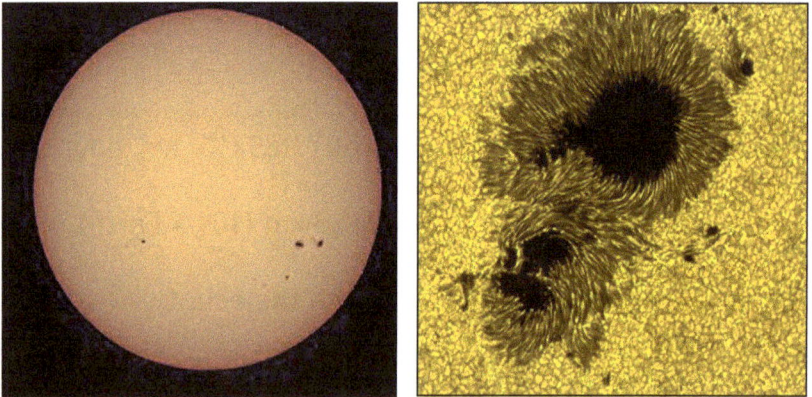

Figures 5.11 & 5.12: Several sunspots can be seen at lower right. These can be tracked across the face of the Sun to measure its rotation (Photo: NASA)

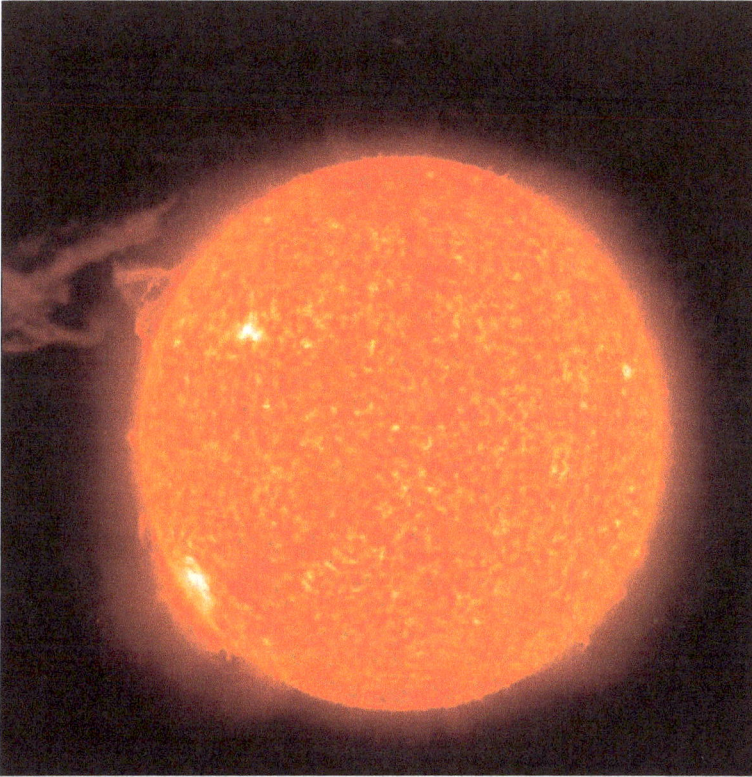

Figure 5.13: Showing the granular nature of the Sun's surface and several sunspots with dark centres (umbra) surrounded by lighter penumbra (Photo: NASA )

- temperatures from about 15 million degrees Celsius in the central core to $5,500^{0}C$ on the surface

- composition (by quantitative spectral analysis) of:

73.46% hydrogen
24.85% helium
0.77% oxygen
0.29% carbon
0.16% iron
0.12% neon
0.09% nitrogen
0.07% silicon
0.05% magnesium
0.04% sulfur

plus smaller traces of the other
92 natural elements.

- natural satellites of 8 major planets (Mercury, Venus, Earth, Mars, Jupiter, Saturn, Uranus, Neptune), many minor (so-called dwarf) planets including Pluto, Eris, Makemake, Huamea, Quaoar, Sedna and Ceres as well as countless asteroids, comets and minor debris.

- star classification (Spectral Class) G-type Main Sequence Star (G2V)

- luminosity or the total amount of energy emitted by an astronomical object such as a star per unit time which is related to the brightness is $3.846 \times 10^{26}$ watts (W). Values for luminosity of other stars are often given relative to the Sun ($L_\odot = 1$).

- apparent magnitude (*m*) of −26.74, which is a measure of the brightness seen from Earth, adjusted to the value it would have in the absence of the atmosphere. The brighter an object appears, the lower its magnitude value and this scale is logarithmic, that is based on powers of 10 so a difference of one in magnitude equals a change in brightness by a factor of about 2.5.

- absolute magnitude of 4.83 which is the hypothetical apparent magnitude of an object if it were placed and seen at a standard distance of exactly 10.0 parsecs (or arc seconds) or 32.6 light years. A **Light Year** (LY) is the distance that light travels in one year at a velocity of $3.00 \times 10^8$ m/s or about of $9.4607 \times 10^{12}$ km).

After an age of 4.5 billion years, this middle-aged star will probably follow the same pattern as most **main-sequence stars** and will remain stable for about another 5.5 billion years, by which time its hydrogen fusion in the core will have stopped and it would have contracted under its own gravity. This contraction brings additional hydrogen into a zone where the temperature and pressure are adequate to cause fusion in a shell around the core. This produces higher temperatures and so the outer layers of the star then expand greatly against gravity, spreading over a much larger surface area, resulting in a lower surface temperature. This changes the emitted light colour towards the red end of the visible spectrum and the Sun becomes a **red giant** star.

## 5.3 The Major Planets

Since the beginnings of astronomy, humankind has noticed that some of the stars (a term generally used in older times for everything seen in the sky) seemed to move independently to all of the others which moved across the sky from east to west, many as regular patterns or groups of stars or **constellations**. These objects the Greek astronomers called planets, from their names *plánēs aster*, meaning wandering star. The planets generally move across the sky from east to west also but at times they also appear to make loops across the sky. This apparent **retrograde motion** or the reversed direction, of the planets is due to the lapping of the Earth as it overtakes or is overtaken by the other planets in their orbits, much the same as an athlete on a circular running track will notice other athletes appearing to run backwards as he/she is overtakes them. Mercury and Venus will only ever appear at low angles because they are closer to the Sun and one must look in towards the Sun, near sunset or sunrise, to see these planets. Because Mercury is in closer to the Sun than Earth, it is termed an **inferior planet**, it never appears to move far from the Sun and its elongation or the angle between the Sun and the planet, reaches a maximum of no more than 28°. Similarly Venus has an elongation of between 45° and 47°. Planets further out than Earth, the **superior planets**, especially Jupiter and Saturn, will appear to move very slowly because of their great distances from Earth. There are a number of terms used in the positional relationships between the planets and the Sun. The most common are:

- **Conjunction** which is the alignment of two celestial bodies as seen from the Earth with reference to the Sun. When a body is in conjunction with the Sun, it is in line with the Sun and so cannot be seen from Earth. Mercury and Venus, as inferior planets, have two positions of conjunction - when they are directly between the Earth and the Sun it is in inferior conjunction, but when either is on the far side of the Sun from the Earth, then they are in superior conjunction. When at inferior conjunction, Mercury and Venus are seen as outlines as they transits across the face of the Sun.

- **Elongation** which is the angle between the Sun and a planet with the Earth as the reference point. The greatest elongation of an inferior planet, such as Mercury and Venus, occurs when it is at a tangent ($90^0$) to the observer on Earth. When it is visible after sunset, it is at its greatest eastern elongation or directly east of the Sun and so can be seen setting after it. When it is visible before sunrise, it is at its greatest western elongation being west of the Sun and so is seen before the Sun rises. When a planet is at its greatest elongation it is able to give the maximum reflection of light from the Sun so it is at its best position for viewing from the Earth. The value of the greatest elongation (west or east), for Mercury, is between 18° and 28°; and for Venus between 45° and 47°. This value varies because the orbits of the planets are elliptical, rather than perfect circles and that the orbital planes, their flat surfaces of their orbits around the Sun, are not exactly on the same level.

- **Quadrature** occurs when a celestial object makes a right angle with respect to the direction of the Sun. It is most useful when describing the apparent position of a superior planet such as Mars, or the Moon at first and last quarters.

- **Opposition** occurs when two celestial bodies are on opposite sides of the sky, viewed from Earth and only concerns the superior planets.

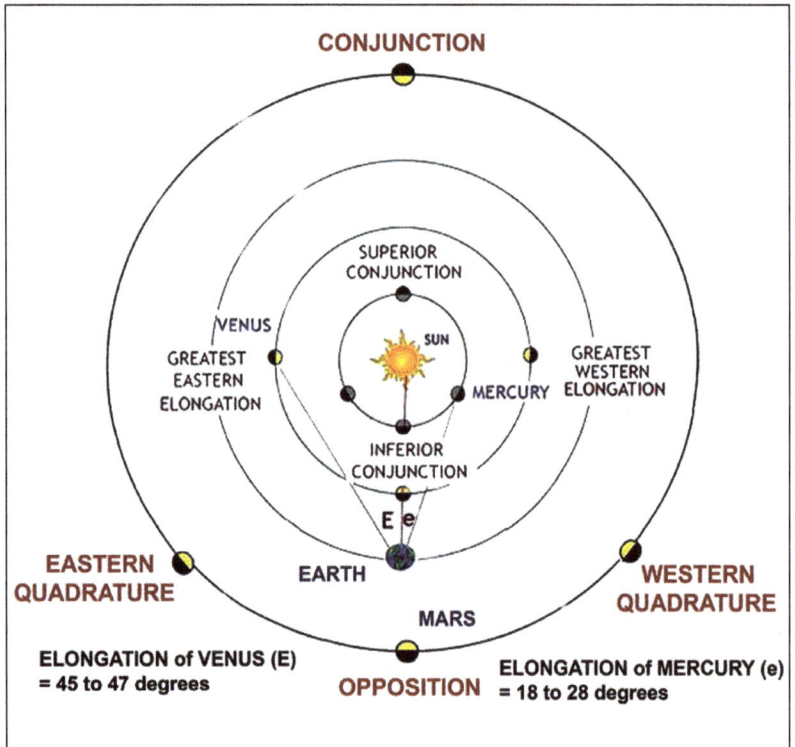

Figure 5.14 Diagram showing positional terms for the Sun, Earth and inner planets

All planets, the Sun, the Moon and the main constellations of the zodiac seem to be on the same pathway through the sky called the ecliptic. It is so named because eclipses of the Sun and Moon appear along this pathway. The position of the ecliptic is due to the comparative viewpoint of the stars, the Sun and other Solar System bodies as seen from the Earth as it rotates on its axis. It is therefore very useful for an observer to know the pathway of the ecliptic across the sky because this will be the pathway taken by the planets as well as the Moon and Sun. When viewed with the naked eye, planets, as reflectors of light rather than luminous sources of light, maintain a steady image without the twinkling effect common to stars. This is because the stars are so far away, that their light acts as though it is coming from a point source and so it is refracted more in the Earth's moving atmosphere. Light from planets, which are closer and acting like a wider source, is not refracted as much and so, with good eyesight and certainly a low-powdered telescope, they will show that they are a disk.

Much of recent knowledge about the planets and other constituents of the Solar System has come from unmanned spacecraft as  flybys, orbiters, static landers and probes and some small rovers landed on Venus (the Soviet's *Venera 7* to *16* programs) and  Mars (NASA's *Sojourner, Spirit, Opportunity & Curiosity*). There have also been several successful contacts made with asteroids, such as Japan's Huyabusa Program which returned samples from the asteroid Itokawa in 2010, and comets such as the European Space Agency's (ESA) *Rosetta* and *Philae* missions. Missions to the outer planets, Saturn, Jupiter, Uranus and Neptune have mostly been by flyby missions, although Jupiter has been visited by the *Galileo Orbiter* and *Probe* missions of NASA with the German space agency in 1995, the *Cassini* spacecraft of NASA-ESA-Italian space agency (ASI) which went into orbit around Saturn in 2004, and the *Huygens*

*Spacecraft* of the ESA which landed on Saturn's moon, Titan. *Pioneer 10 & 11, Voyager 1 & 2* and the *New Horizons* programs, all from NASA, have explored the furthest planets and beyond since 1974.


## 5.4 Mercury (Symbol: ☿ - The Roman God's Helmet and Staff)

Figure 5.15: False coloured mosaic image of Mercury from data from the *Messenger* spacecraft (Photo: NASA)

Mercury is the closest planet to the Sun and named after the Roman messenger god (the Greek god: Hermes) who was fast-moving and often invisible; very much like the planet which is difficult to see because it is close to the Sun. Because of this, Mercury can only be seen as the Sun rises in the morning or as it sets in the evening, but not in the middle of the night. It is seen only when it is between 18° and 28° elongation above the horizon. Like the Moon, it also shows a complete range of phases as it moves around its orbit relative to Earth.

Some of the main facts about Mercury are:

- average distance from the Sun of 57.9 million km (0.39 astronomical units AU)

- orbital period (year) 87.969 Earth days

- average orbital velocity of 47.362 km/ sec

- sidereal rotation period (day) of  58.646 days

- equatorial rotation velocity of 10.892 km/hour

- average radius of 2439.7 km

- axis of rotation is very small, being only tilted at 2.11 degrees

- surface area of  $7.48 \times 10^7$ km$^2$ (0.147 of  Earth's)
- volume of  $6.083 \times 10^{10}$ km$^3$ (0.056 of Earth's)

- mass of  $3.3011 \times 10^{23}$ kg ( 0.055 of Earth's)

- average density of  5.427 g/cm$^3$

- gravitational acceleration of 3.7 m/s$^2$

- escape velocity 4.25 km/sec

- active tectonism some evidence of very old volcanism

- magnetic field is very weak, being only about 1.1% as strong as Earth's magnetic field

- surface temperatures have the greatest variation because Mercury has almost no atmosphere to retain heat, with temperatures ranging from −173 °C at night to 427 °C during the day at some equatorial regions. The poles are constantly below −93 °C

- air pressure and composition - only a very small trace of an atmosphere has been detected and this consists of:

| | |
|---|---|
| oxygen | 42.0% |
| sodium vapour | 29.0% |
| hydrogen | 22.0% |
| helium | 6.0% |
| potassium vapour | 0.5% |

also there are very small traces of argon, nitrogen, carbon dioxide, water vapour (some water ice has been found in permanent shadow areas at Mercury's north pole) xenon, krypton and neon. No natural satellites.

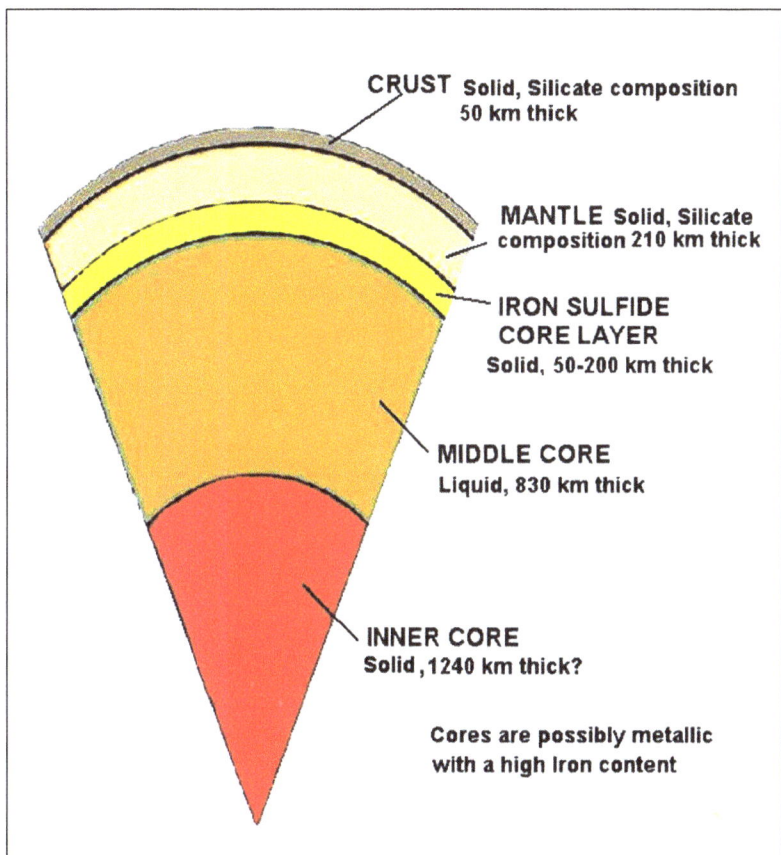

Figure 5.16: Diagram showing Mercury's possible interior (data from the *Messenger* spacecraft)

## 5.5 Venus (Symbol: ♀ The Mirror of the Roman Goddess of Love And Beauty)

Figure 5.17: A cloud-covered Venus seen from the *Pioneer* orbiter in 1978 (Photo: NASA)

Venus is the second planet from the Sun and after the Moon, it is the brightest object in the night sky, bright enough to cast shadows. Because Venus is an inferior planet, it never appears to venture far from the Sun having an elongation between 45° and 47°. Because of this it called both the Morning Star and the Evening Star for its time of brightest appearance.

Venus is a **terrestrial planet,** and is similar to the Earth in size, mass, proximity to the Sun and bulk composition. However, in other ways it is completely different. It has the densest atmosphere of the four terrestrial planets mostly of carbon dioxide (96%), with reflective clouds of sulfuric acid and an atmospheric pressure 92 times that of Earth's. Because of the thick atmosphere acting like a glasshouse, and being closer to the Sun, Venus' surface temperature is about 462°C and is the hottest planet in the Solar System. There seem to be very few craters on the surface due to the thick atmosphere and its surface is dominated by many huge shield volcanoes and basaltic lava plains.

Some of the main facts about Venus are:

- average distance from the Sun of 108.2 million km (0.723 astronomical units AU)

- orbital period (year) of 224.701 days

- average orbital velocity of 35.02 km/sec

- sidereal rotation period (day) of 243.025 days in the opposite direction to the rotation of the other planets (i.e. it has retrograde rotation). It was once thought that this may have been caused by a glancing blow from a passing body in the early days of the Solar System, but current theories suggest that the dense atmosphere, its proximity to the Sun and the relatively large core may have had a tidal effect on the whole planet causing it to slow and reverse direction. The exceptionally long days may also be caused by this effect.

- equatorial rotation velocity of 6.52 km/hour

- average radius of   6051.8 km (0.9499 of Earth's)

- axis of rotation of 2.64° retrograde

- surface area of   $4.6023 \times 10^8$ km$^2$ ( 0.902  Earth's)

- volume of   $9.2843 \times 10^{11}$ km$^3$ (0.866 Earth's)

- mass of   $4.8675 \times 10^{24}$ kg (0.815 Earth's)

- average density of  5.243 g/cm$^3$

- gravitational acceleration of 8.87 m/s$^2$

- escape velocity 10.36 km/s

- active tectonism – the surface was finally mapped in detail by the *Magellan* orbiter in 1990-91 and the ground shows evidence of extensive volcanism with many larger volcanoes than on Earth. The large amount of sulfur in the atmosphere suggests that there have been some recent eruptions.
- magnetic field none detected, probably because of its slow rotation and large solid component of the core

- surface temperatures vary little on this uniformly hot planet with surface temperatures around 462 °C

- air pressure and composition is extremely high at about 92 bars (i.e. 92 times that of Earth with a composition of:

|                |                        |
|----------------|------------------------|
| carbon dioxide | 96.5 %                 |
| nitrogen       | 3.5 %                  |
| sulfur dioxide | 0.05% (giving clouds of sulfuric acid when combined with water vapour) |

with traces of argon, water vapour, carbon monoxide, helium, neon, carbonyl sulfide and hydrogen halides.

- no natural satellites.

Figure 5.18: The north polar region of Venus seen using RADAR from the *Magellan* orbiter to see through the atmosphere (Photo: NASA)

Figure 5.19: A vertically-exaggerated radar view of the 8 km high
volcano Maat Mons taken from the *Magellan* orbiter (Photo: NASA)

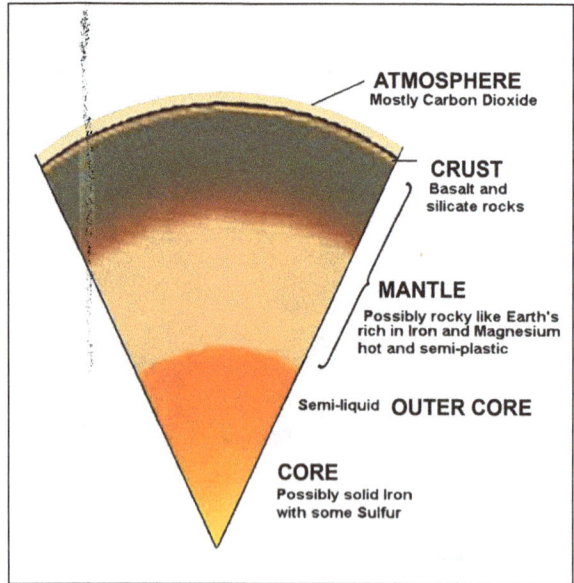

Figure 5.20: Diagram showing a possible view of the interior of Venus

## 5.6 Mars (Symbol: ♂ - The Spear and Shield of the Roman God of War)

Figure 5.21: Mars showing Olympus Mons (top left), the three volcanoes (from top) Ascraeus Mons, Pavonis Mons, and Arsia Mons and the Valles Marineris (centre right) (photo: NASA).

The fourth planet from the Sun, Mars is the second smallest planet in the Solar System (after Mercury) and is often referred to as the Red Planet because of its colour in the night sky. This reddish colour is due to the abundance of iron oxides (mostly haematite) on its surface. Mars has a very thin atmosphere and has surface features not unlike the drier parts of Earth with deserts, volcanoes, valleys and a polar ice cap.

The rotational period of a little over 24 hours and its similar axial tilt produces seasons very similar to Earth. It appears to travel through the night sky in a series of loops, explained by the 16th Century astronomer, **Tycho Brahe** (Danish: 1546-1601), who believed that Mars orbited the Earth, as **epicycles**, or loops within the circle of this orbit. Later astronomers who followed the heliocentric model were able to show that these loops where due to the relative rotational velocities of the Earth and Mars as Earth overlapped Mars in their separate orbits around **the Sun.**

**Figure 5.22: Multi-exposure of Mars in the night sky showing its retrograde (reversed) Motion**

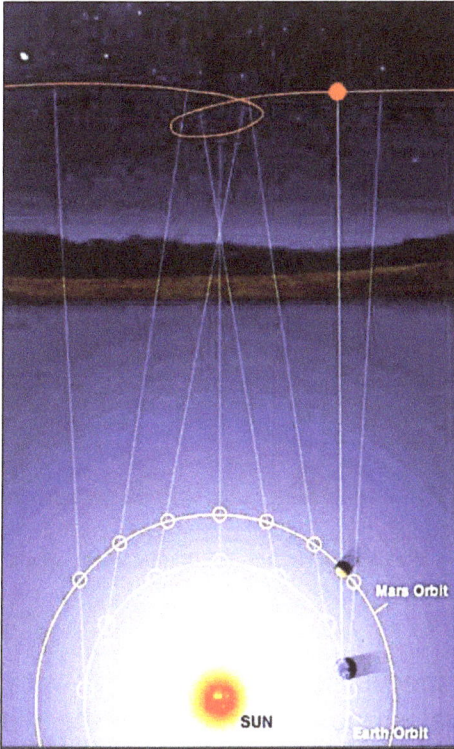

Figure 5.23: Diagram showing how this apparent reversal occurs as the Earth overtakes and passes Mars in orbit (Photo: NASA)

Its surface features are dramatic, with many large shield volcanoes including the huge Olympus Mons, the largest volcano known in the Solar System and the Valles Marineris, a huge canyon running for over 4000 km, 200 km wide and over 7 km. deep - one of the largest canyons in the Solar System. In contrast, in the northern hemisphere is the smooth Borealis Basin which covers 40% of the planet and may be a giant impact feature. On 28 September 2015, NASA announced the presence of briny flowing salt water on the Martian surface.

Figure 5.24: Closer view of the Valles Marineris, the 4000 km canyon on Mar's surface (Photo: NASA/JPL)

There had been much speculation about surface water on Mars. Telescopic observation noted that the northern polar cap often increased during the Martian winter and decreased during summer and long, dark striations, described by the Italian astronomer **Giovanni Schiaparelli** (Italian: 1835 -1910) in 1877 as *canali* or channels also seemed to change. Unfortunately, this term was wrongly translated into English as canals causing much speculation about the possibility of inhabitants of Mars. Improved astronomical observations in the 20th Century revealed that these canals were, in fact an optical illusion. Later, modern high resolution mapping of the Martian surface by spacecraft shows no such canals, but there was good geological evidence gathered by unmanned Mars missions to suggest that Mars once had large-scale water drainage on its surface at some earlier time. In 2005, radar data revealed the presence of large quantities of water ice at the poles as well as at mid-latitudes. The Mars rover *Spirit* sampled chemical compounds containing water molecules in March 2007 and the *Phoenix* lander directly sampled water ice in shallow Martian soil on July 31, 2008.

Figure 5.25: Dark narrow streaks (called Recurring Slope Linae) down the walls of Garni Crater on Mars. (NASA/JPL/University of Arizona

Figure 5.26: View taken by the *Curiosity* rover of what appears to be cracked lake sediments (Photo: NASA/JPL-Caltech/MSSS)

Some of the main facts about Mars are:

- average distance from the Sun of  227.9 million km

- average distance from the Earth of  from 55.7 to 401.3 million km

- orbital period (year) of  686.971 days

- average orbital velocity  24.077 km/sec

- sidereal rotation period (day) of 24 hours 37 minutes and 22 seconds. the Martian day has been given a name – the Sol

- equatorial rotation velocity of 868.22 km/hour

- average radius of 3389.5 km

- axis of rotation of 25.19°

- surface area of 144,798,500 km$^2$ (0.284 Earth's)

- volume of 1.6318×10$^{11}$ km$^3$ (0.151 Earth's)

- mass of 6.4171×10$^{23}$ kg (0.107 Earth's)

- average density of 3.9335 g/cm$^3$

- gravitational acceleration of 3.711 m/s$^2$

- escape velocity of 5.027 km/s

- magnetic field and tectonism no evidence of a present-day magnetic field has been detected, although there is some evidence from rock samples analyzed by the Martian rovers that some parts of the planet's crust have been magnetized, and that as in Earth-based basalts, there has been some alternating polarity reversals of its field in the past. This paleomagnetism of magnetic minerals is similar to what is found on Earth and one theory suggests that these bands may be due to plate tectonics on Mars four billion years ago, before the planet's magnetic field ceased. There is, however, some evidence from sharp images from the Olympus Mons region that there may have been some relatively recent volcanic eruptions on Mars over the last two to three million years.

- surface temperatures vary from −143 °C near the poles to about 35 °C at the equator

- air pressure is very low compared to that on Earth at about 0.636 kPa (Earth's pressure is 101.325 kPa). the Martian atmosphere is composed of:

| | |
|---|---|
| carbon dioxide | 95.97% |
| argon | 1.93% |
| nitrogen | 1.89 |
| carbon monoxide | 0.0557% |

and very small traces of water vapour, neon, krypton, xenon and methane.

- natural satellites. Mars has two, very small and irregularly-shaped moons, Phobos (horror) and Deimos (terror) which both measure only in tens of kilometres and are heavily cratered. Their close proximity and rapid orbits suggest that they may have been captured asteroids.

Figure 5.27: Phobos, the largest moon of Mars orbits the planet every 7 hours and 39 minutes at a low altitude of 6000 km. It has dimensions of only 27 × 22 × 18 km. (Photo: NASA from the Mars *Reconnaissance* orbiter)

Mars is perhaps the only other planet where humankind could live, albeit with some difficulty. At its closest point, it is still some 55 million kilometres from Earth and it would take considerable time, perhaps 6 to 9 months, for a spacecraft to get there from Earth. This would probably be achieved by launching several cargo rockets into Earth orbit where the main Mars vessel would be constructed. Perhaps several may be needed as the total payload required for such a trip would be very large. Once complete, the vehicle(s) would be accelerated towards Mars on a complex trajectory which would bring them to Mars orbit in the quickest possible time. After preliminary orbital reconnaissance, the landing team would then descend, probably using smaller craft attached to the main vehicle, to the surface where habitats and other buildings would be constructed. Several descents would be needed to take the supplies to the Martian surface even if the buildings were light and prefabricated for easy assembly. Subsurface dwelling in caves or covered excavations may be an answer to some of the environmental problems. Even allowing for a short exploration of the immediate vicinity of the landing, the total time away from Earth would probably be one to two years – more if extended exploration is needed. Some comparison between the Mars environment and that of Earth is shown in the table on the next page:

| CONDITIONS | ON EARTH | ON MARS | NOTES |
|---|---|---|---|
| Day-Night | 24 hours- 12 of each | About the same | Colder nights |
| Gravitation | Grav. Accel. g = 9.8 m/s² | About 1 third g =3.711 m/s² | Lifting easier but muscle wasting |
| Terrain | Variable | Similar to a dry rocky Earth | Similar to rocky deserts |
| Water | Available known sources | Yes but sources unknown | Careful choice of landing site near a known water source |
| Food | Available can be grown | Needs to be transported and condensed | Gardens on Mars? |
| Power | Available grid | No natural sources | Compact. Solar/nuclear |
| Atmosphere | Suitable for life | Thin, unsuitable | Air supply from water by solar electrolysis? |
| Temperatures | Suitable for life | Colder and varies rapidly | Can be overcome |
| Radiation | Atmospheric shielding | No shielding from ultra violet, cosmic rays, solar wind | Suits and habitat suitable |
| Meteorites | Atmospheric shielding | Considered minimal | Warning watch needed |
| Weather | Livable conditions | Dust storms | Monitored from orbit |
| Seasons | Four distinct seasons | Also has seasons - but longer | No major problem |
| Communications | Quick and good variety | Time delays (15 min. one way | Set times for sending/receiving |

| | | delay) | |
|---|---|---|---|
| Social & Psychological | Considerable support | Isolation of a small group | Good selection of crew needed |
| Bio Hazard | Generally known | No resistance to any Mars germs | Isolation in suits, habitat. Medical facilities needed |
| Wastes | Wider choice for removal | Recycling and storage required | Limited with threat of contamination |

Table 5.1: Comparison of conditions on Mars and Earth

Going to Mars has several problems. These include:

- Huge payloads needed for the voyage would demand a very big spacecraft which would have to be built in Earth orbit.

- Food would have to be transported for a two year (plus) mission unless it can be supplemented with minimal water, indoor greenhouse hydroponic system.

- Water has been found on mars but further detailed locations need to be identified and a suitable landing site chosen near a water source.

- Huge expense in building, launching and maintaining the mars vehicles is the current problem. whilst exploration and colonization of space is a human desire, financing such enterprise still relies on mainly government funding.

- Travelling through the space environment for very long periods is now a matter of time rather than ability to do so. Experimentation aboard the various space stations around the Earth has shown that this can be done.

- Complexities of the trajectory would require precise stellar navigation so as to reach mars when it is at its closest point to Earth. This too, is a problem with known solutions; only the large distance, delay in communications and control during such a long time are the new factors.

- Orbiting and the transfer of crew and material from main vehicle to shuttles to and from the surface should not be a major problem, as these have been exercised in Earth orbit.

- Construction of the habitats under reduced gravity would be easier than on Earth, especially using suitable small machines, but there is little margin for error.

- Daily work tasks outdoors would be in special suits required to provide respiration, good body temperature and radiation shielding. This should be overcome by drawing upon the experiences from the Apollo Missions, but there would still be limitations in outdoor work and exploration requiring an enclosed, large roving vehicle to cover distance and a stable support base.

Figure 5.28: Artist's impression of Mars Astronauts working on the surface (Photo: NASA/JSC)

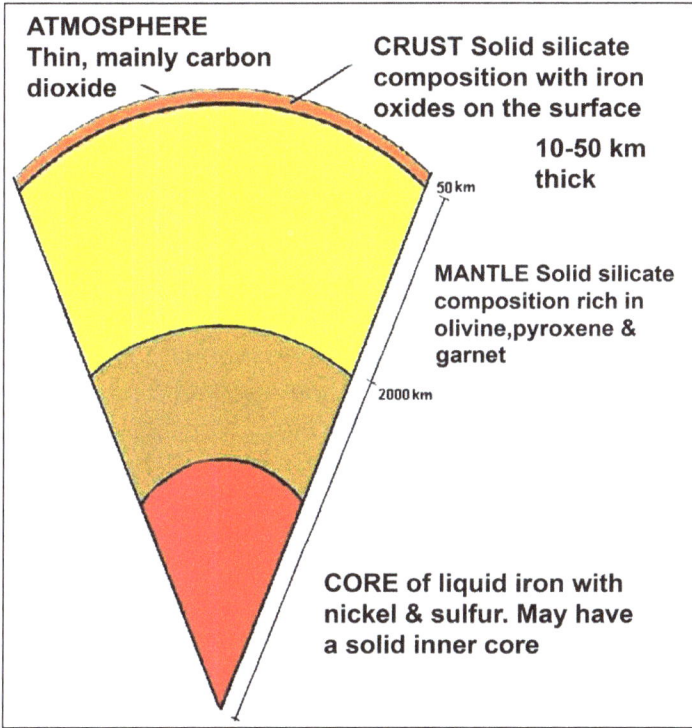

**ATMOSPHERE**
Thin, mainly carbon dioxide

**CRUST** Solid silicate composition with iron oxides on the surface

**10-50 km thick**

50 km

**MANTLE** Solid silicate composition rich in olivine,pyroxene & garnet

2000 km

**CORE** of liquid iron with nickel & sulfur. May have a solid inner core

Figure 5.29: Diagram showing the possible interior of Mars

## 5.7 A Big Gap

By the 18<sup>th</sup> Century, telescopes had developed to a degree which enabled considerable study and measurement of all of the six known planets – Mercury, Venus, (Earth), Mars, Jupiter and Saturn. Distances from these planets and the Sun were known, although in astronomical units and conversion to actual working units of distance was one of the major directions of astronomy. In a world where the ancient Greek idea that nature was a matter of perfect geometry, and where new advances in mathematics was an obsession, the observation that there was a considerable distance between Mars and the next planet out from the Sun, Jupiter begged further thought.

In 1766, **Johann Daniel Titius** (German: 1729 -1796) noticed that there seemed to be a mathematical series relationship between the distance of the planets from the Sun and he used this series to predict that another, as yet unknown planet should be found about 2.8 AU from the Sun. In 1772, **Johann Elert Bode** (German: 1747 - 1826), a young astronomer mentioned Titus' prediction in his own work and went on to elaborate the mathematical series thus:

"Let the distance from the Sun to Saturn be taken as 100, then Mercury is separated by 4 such parts from the Sun. Venus is 4+3=7. The Earth is 4+6=10. Mars is 4+12=16. Now comes a gap in this so orderly progression. After Mars there follows a space of 4+24=28 parts, in which no planet has yet been seen. From here we come to the distance of Jupiter by 4+48=52 parts, and finally to that of Saturn by 4+96=100 parts."

Applying this mathematical progression gave a reasonable (for that time) approximation for the known planets:

| PLANET | MATHEMATICAL CALCULATION | PROPOSED (ACTUAL VALUE) |
|---|---|---|
| Mercury | start at 4 and divide by 10 = | 0.4 A.U. approx. (0.39 A.U.) |
| Venus | 4 + 3 = 7 divide by 10 = | 0.7 (0.72 A.U.) |
| Earth | 4 + 6* = 10 divide by 10 = | 1.0 A.U standard |
| Mars | 4 + 12* = 16 divide by 10 = | 1.6 (1.52 AU) |
| No Planet Yet? | 4 + 24* = 28 divide by 10 = | Should be at 2.8 A.U. (Ceres, discovered in 1801 is at 2.6 A.U.) |
| Jupiter | 4 + 48 = 52* divide by 10 = | 5.2 A.U. (5.20 A.U.) |
| Saturn | 4 + 96 = 100* divide by 10 = | 10 A.U. (9.54 A.U.) |

Table 5.2: Titus-Bode predictions for the planets known at that time
*note that the number added to the base of 4 is being doubled for the each planet

Beyond Jupiter, the last of the planets out from the Sun at that time, the next planet should be at:

(unknown)   4 + 192 = 196 (or 19.6 A.U.).

When originally published, this theoretical, mathematical relationship, worked well for all the then-known planets with a gap between the fourth and fifth planets. It was regarded as interesting, but of no great importance until **Sir**

**William Herschel** (German-English: 1738 – 1822) discovered Uranus in 1781 exactly at the predicted 19.2 A.U.

Based on this discovery, Bode urged a search for a fifth planet. This came in 1801 when **Guiseppe Piazzi** (Italian: 1746 – 1826) discovered the dwarf planet Ceres about the 2.8 A.U. position between Mars and Jupiter. This mathematical method of astronomical prediction was called the Titus-Bode law and seemed to show the perfection in mathematical modelling of the Solar System.

Whilst this Law was very useful in discovering Uranus and Ceres, it did not work for the next planet to be discovered, Neptune. This was discovered in 1846 by **Urbain le Verrier** (French: 1811 – 1877) and was found to have a distance of 30.1 A.U., not the predicted 38.8 A.U. needed to satisfy the Titus-Bode Law. Also, further observations towards Ceres showed that it was only the biggest of a number of objects, now known to be asteroids in their own diffuse orbit around the Sun between Mars and Jupiter. The Titus-Bode Law fell out of scientific thought as just the luck of some mathematical manipulation.

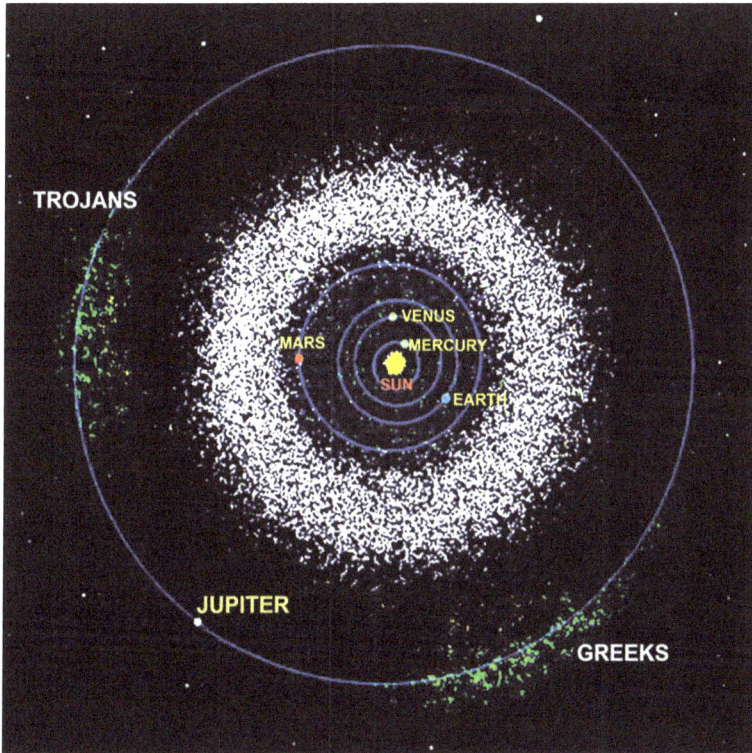

Figure 5.30: Diagram showing the main locations of the asteroids

The **asteroid belt** consists of a very large number of irregularly shaped bodies which orbit between Mars and Jupiter. It is often called the Main Asteroid Belt to distinguish it from other asteroid populations in the Solar System such as those close to Earth, the Near Earth Asteroids, and those which are located within the orbits of the other planets following them in their path. These asteroids which lie within the orbits of planets are called the **Trojan asteroids**, as they remained hidden in the orbits of these planets much as the Ancient Greeks remained hidden in the Trojan Horse of Greek mythology. Such asteroids accumulate at the Lagrange Points within a planet's orbit where the gravitational pull of the Sun and the planet exactly balances and allows the centrifugal force of the smaller bodies to maintain a stable orbit. The larger asteroids are given the names of minor goddess suitable to system of using Roman mythology to name the planets.

About half the mass of the main belt is contained in the four largest asteroids: Ceres, Vesta, Pallas, and Hygiea, but the combined masses of all in the asteroid belt is approximately only about 4% that of the Moon. Ceres, the only body to rate the status of a dwarf planet, is only about 950 km in diameter, whereas Vesta, Pallas, and Hygiea have average diameters of less than 600 km. The remaining asteroids range down to dust particle size and it is so thinly distributed that numerous unmanned spacecraft have passed through it without incident. Recent studies have shown that this Main belt has an inner zone of rocky material and an outer zone of icy material. The **Nice hypothesis** (named after the French city and pronounced *neece*) suggests that the inner zone of rock was formed in place and when heated by the Sun, much of the ice and other volatile material was driven off. The outer zone is then thought to have been pushed inwards from the deeper, colder regions of the early Solar System during the migration of the Gas Giants outwards.

## 5.8 Jupiter (Symbol: ♃ Jupiter's Thunderbolt or His Eagle)

Figure 5.31: The planet Jupiter seen from NASA's Cassini-Huygens spacecraft en route to Jupiter in 2000. Note the Giant Red Spot (lower right) and the turbulence within the many gas bands (Photo: NASA).

Jupiter is the fifth and largest planet in the Solar System, so it is fitting that it is named after the Romans' chief god. In the night sky it can appear as a very bright disk and is one of the brightest objects to be seen.

Jupiter has a mass one-thousandth that of the Sun, but two and a half times that of all the other planets in the Solar System combined. Jupiter is a **gas giant** as it is mainly composed of hydrogen with a little helium, but it may also have a rocky core of heavier elements but lacks a well-defined solid surface. It has a very rapid rotation, giving the planet the shape of a flattened sphere with an equatorial bulge making it an oblate spheroid. The outer atmosphere has a number of distinct gas bands which rotate, resulting in considerable turbulence and storms along their interacting boundaries. One of these storms is the **Great Red Spot**, which has observed since the early days of the telescope in the 17th century. In more recent times, Jupiter has been explored by several unmanned flyby missions such as *Pioneer, Voyager, New Horizons,* and the *Cassini-Huygens* Missions, and the *Galileo Orbiter* and entry *Probe*. The *Juno Orbiter* was launched in 2011 and successfully went into orbit in July 2016. Future targets for exploration in the Jupiter system include the probable ice-covered liquid ocean of its moon Europa. Jupiter also has a strong magnetic field.

The Giant Red Spot is a very large high pressure or anticyclonic storm system which has changed in size only slightly over the four hundred years of its observation from Earth. This oval object rotates with a period of about six Earth days and its dimensions vary from 24-40,000 km west to east and from 12-14,000 km south to north  and is large enough to contain two or three Earths. Infra-red data suggests that the Giant Red Spot is colder than the surrounding cloud belts, which are composed of icy methane at speeds over 500 km/hour, and so is probably higher in altitude. Some computer modeling suggests that in the upper atmosphere, the hydrogen and helium gases have become more metallic and electrically charged. The immense heat at the core of the planet drives gaseous hydrogen and helium upwards in bubbles, as convection currents. As these bubbles break through the layer of metallic hydrogen and helium, they spread out horizontally and under the influence of planetary rotation, they also rotate in an anticlockwise direction.

Some of the main facts about Jupiter are:

- average distance from the Sun of  778.3 million km (5.203 au)

- orbital period (year) of 11.862 Earth years

- average orbital velocity of 13.07 km/sec

- sidereal rotation period (day) of 9.84 Earth hours

- equatorial rotation velocity of 45300 km/hour

- equatorial diameter of 142,984 km and polar diameter of 133,708 km (average diameter approximately 11 times that of Earth)

- axis of rotation is tilted at 1.303° to the ecliptic

- surface area of  $6.1419 \times 10^{10}$ km$^2$ (121.9 Earth's)

- volume of $1.4313 \times 10^{15}$ km$^3$ (1321.3 Earth's)

- mass of $1.8986 \times 10^{27}$ kg (317.8 of that of Earth and 1/1047 that of the Sun)

- average density of 1.326 g/cm$^3$

- gravitational acceleration of 24.79 m/s$^2$

- escape velocity of 59.5 km/sec

- active tectonism not observed due to deep cloud cover

- magnetic field which is generated by electrical currents in the planet's outer core of liquid metallic hydrogen and its **magnetosphere** (the cavity created in the solar wind by the planet's magnetic field) is the largest of any planet and extends up to seven million kilometres in the Sun's direction and almost to the orbit of Saturn in the opposite direction.

- surface temperature is about 120 k (-153$^0$C)

- atmospheric pressure in the cloud layer is very high (> than 1000 bars) with a composition of:

| | | |
|---|---|---|
| hydrogen | 89.8 | % |
| helium | 10.2 | % |
| methane | 0.3 | % |
| ammonia | 0.026 | % |

and slight traces of other gases such as ethane and water vapour with ices of ammonia, water and ammonium hydrosulfide

- natural satellites consist of 67 moons (18 named), many other smaller objects and several very faint rings having ice and dust particles

The giant planet also has a faint ring system orbiting it, and at least 67 moons, including the four largest, Io, Europa, Ganymede, and Callisto. These are called the Galilean Moons as they were first seen by Galileo with his telescope in 1610, and named after the many lovers of this Roman god. Ganymede, the largest of these moons, has a diameter greater than that of the dwarf planet Mercury, and Europa is of great interest because of its ice covering.

Some interesting details about these moons are:

- Io is the innermost moon and slightly larger than the Earth's Moon with an average radius of 1,821.3 km - about 5% larger than Earth's Moon. It has over 400 active volcanoes, mostly erupting sulfur, as a result of the tidal effects of Jupiter's gravitational pull on the moon's surface. It takes Io about 42.5 hours to complete one orbit around Jupiter, keeping one face pointed toward Jupiter. It probably has an internal structure similar to the terrestrial planets being mostly composed of silicates with an iron core.

- Europa is the sixth closest moon to Jupiter and the smallest of the Galilean Moons with an average diameter of 3121.6 km, and has an orbital period of 3.5 days. It has a very faint atmosphere of mainly oxygen and its surface is marked by ice with many cracks with few craters. The surface seems to have been re-worked into a new, smooth surface and its appearance has led to the hypothesis that a there may be liquid water beneath the surface ice which could possibly have some forms of extra-terrestrial life. This hypothesis proposes that heat from tidal flexing of the moon may allow this water to remain liquid and also drive the tectonic activity. Below the water-ice crust, Europa probably has a silicate rock mantle and an iron-nickel core.

- Ganymede is the seventh satellite outward from Jupiter with an orbital period of 7.1 days. It is the largest in the Solar System, and the only moon known to have a magnetosphere due to a magnetic field caused by a liquid iron core. Its interior has several layers of ice over liquid water which overlays a silicate rock mantle and an iron-rich, liquid core. Its surface is composed of darker regions, heavily cratered (dated to an age of four billion years ago) which covers about a third of the satellite and a slightly younger, lighter region, cut by many grooves and ridges which may be the result of tectonic activity.

- Callisto is the second largest moon of Jupiter with a diameter of 4821 km diameter and is the fourth moon from the planet at a distance which does not experience any tidal effects of Jupiter's gravity. It is, however, tidally-locked to the planet so that the same hemisphere always faces inward Jupiter. Callisto is composed of approximately equal amounts of silicate rock and ices and has a light density of only about 1.83 g/cm$^3$. It probably has a small silicate core surrounded by ice and a sub-surface layer of liquid water. The surface of Callisto is old and heavily cratered.

As well as the large Galilean moons there are four small moons inside their orbit with an average size of about 200km each and six groups of irregularly-shaped small moons orbiting outside of the Galilean group. Some of these smaller moons also have retrograde orbits.

Figure 5.32: Photo montage showing the four Galilean moons next to the giant red spot for size comparison. From the top, the moons are Io, Europa, Ganymede and Callisto. (Photo: NASA)

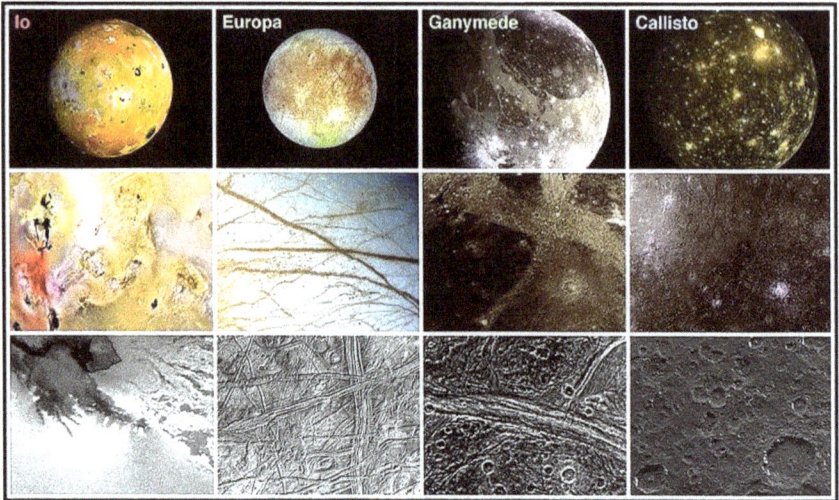

Figure 5.33: The Galilean moons of Jupiter and some of their surface features (Photo: NASA)

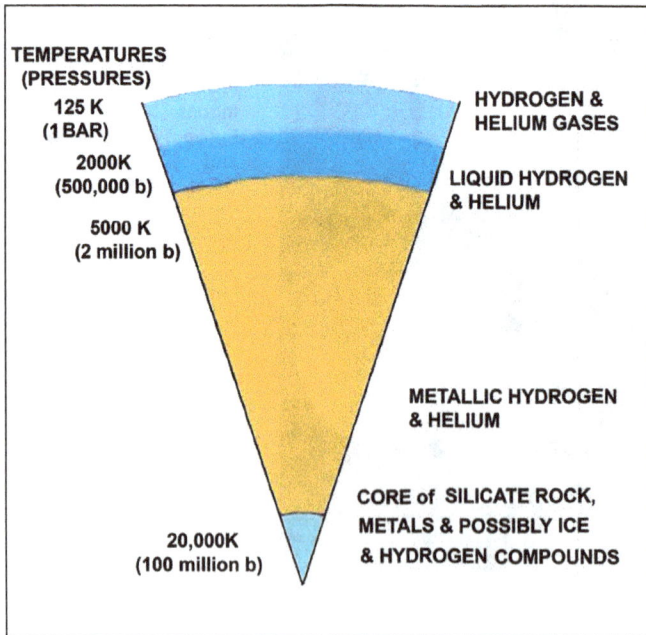

Figure 5.34: Diagram showing the possible interior of Jupiter

## 5.9 Saturn (Symbol: ♄ Represents the Roman god's Sickle)

Figure 5.35: Saturn, the ringed planet as seen from the flyby spacecraft *Cassini* (Photo: NASA)

Saturn is also a Gas Giant and the second-largest in the Solar System with an average radius of 58232 km, nine times that of Earth. It is the sixth planet from the Sun at 1,427.0 million km. (9.54 A.U.) distant. Saturn is named after the Roman god of agriculture and time. In the night sky it can be very bright, but not as much as Jupiter and it has a pale yellow hue due to ammonia crystals in its upper atmosphere. Whilst there are coloured gas bands in its atmosphere, these are not as prominent as in Jupiter. Wind speeds within the atmosphere can reach as much 1,800 km/hour, higher than those on Jupiter. Occasionally, violent, white-coloured storms break through the cloud layers, each one bigger than Earth.

Saturn has a prominent ring system of nine continuous main rings and three discontinuous arcs which are all composed mostly of ice particles. A closer view of the rings showed a series of "spokes" radiating across the ring surfaces. These are believed to be streams of ice particles which have been raised above the levels of the rings. In 2017, the spacecraft *Cassini* performed several complex maneuvers – turning its large dish antennae towards the planet to act as a shield whilst it hurtled through the rings. Having survived this passage, it once more turned so that its antennae faced Earth and resumed radio contact. Initial images showed a very banded upper atmosphere and a large storm cell on the atmosphere's surface.

Figure 5.36: False coloured image of the rings of Saturn taken from the flyby of the spacecraft *Cassini* in 2005 (Photo: NASA)

Figure 5.37: Raw data image of a huge storm in the upper atmosphere of Saturn taken in April 2017 after the spacecraft *Cassini* had passed through the ring system and skimmed the surface of the planet
(Photo: *NASA/JPL - Caltech/Space Science Institute*)

The rings also contain a large amount of rocky debris and dust as well as hundreds of **shepherd moons** within the rings. These clear gaps within the ring and appear to maintain some form of order within the ring system. There are sixty-two other moons orbiting Saturn of which fifty-three have been named. These moons go around the planet on the same orbital plane as the rings. Titan, Saturn's largest moon, and the second-largest in the Solar System, is larger than the planet Mercury, and has a substantial atmosphere. Enceladus, sixth in size of Saturn's moons and only about 500 km. across is mostly covered by fresh, clean ice, reflecting almost all the sunlight, and has a wide range of surface features, including young tectonically deformed areas to older regions with many craters.

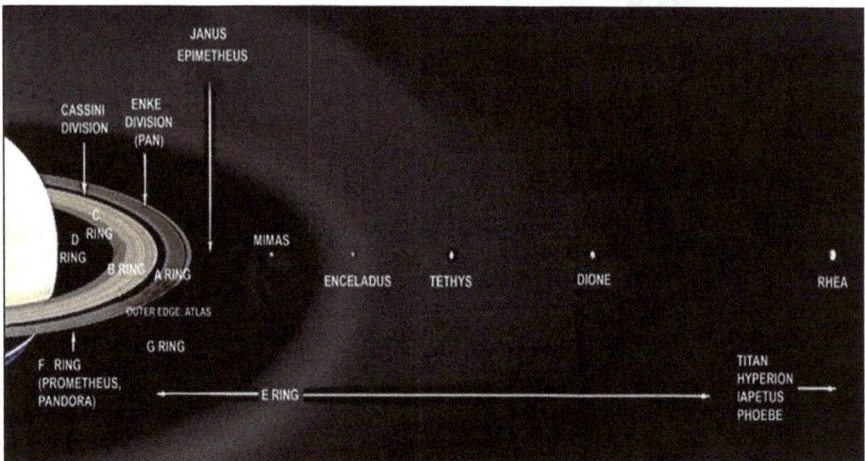

Figure 5.38: Diagram showing Saturn's ring system and some of its major moons

Some of the main facts about Saturn are:

- average distance from the Sun of 1,427.0 million km (9.5 A.U.)

- orbital period (year) of 29.4571 years

- average orbital velocity of 9.69 km/sec

- sidereal rotation period (day) of 10.55 Earth hours

- equatorial rotation velocity of 35500 km/hour

- equatorial diameter of 120,536 km

- axis of rotation is tilted at 2.49° to the ecliptic

- surface area of $4.27 \times 10^{10}$ km$^2$ (83.703 Earths)

- volume of $8.2713 \times 10^{14}$ km$^3$ (763.59 Earths)

- mass of $5.6836 \times 10^{26}$ kg (95.159 Earths)

- average density of 0.687 g/cm$^3$ (this is less than that of water)

- gravitational acceleration of 10.44 m/s$^2$

- escape velocity of 35.5 km/sec

- active tectonism not observed due to deep cloud cover

- magnetic field is quite unlike that of any other planet, showing no tilt to the rotation axis and somewhat smaller than expected at 4.6 x 10$^{18}$ T m$^3$ 580 times larger than that of the Earth

- surface temperatures is about 134 k (–139°C) at the surface

- atmospheric pressure in the upper cloud layer is very high (> 1000 bars) with a composition of:

|  |  |
|---|---|
| hydrogen | 96.0 % |
| helium | 3.0 % |
| methane ($CH_4$) | 0.4 % |
| ammonia ($NH_4$) | 0.01% |

with smaller traces of hydrogen deuteride (a compound of hydrogen and its isotope deuterium), ethane ($C_2H_6$) and ices of ammonia, water and ammonium hydrosulfide ($NH_4SH$).

- natural satellites of over 150 moons (53 named) and smaller moonlets as well as rings of ice, small rocks and dust.

Figure 5.39: Saturn's largest moon Titan has a blurred image due to its atmosphere of nitrogen with methane & ethane (photo: NASA).

156

Figure 5.40: By contrast, Enceladus, at only 500 km across has no atmosphere and is covered with ice (Photo: NASA)

Saturn's interior is probably composed of a core of iron-nickel and silicate rock .This is probably surrounded by a deep layer of metallic hydrogen, followed by layers of liquid hydrogen and liquid helium. Above this and across a transition zone, is a gaseous outer layer of these gases. As with Jupiter, electrical currents in its metallic hydrogen layer are thought to cause its magnetic field, which is weaker than that of Earth.

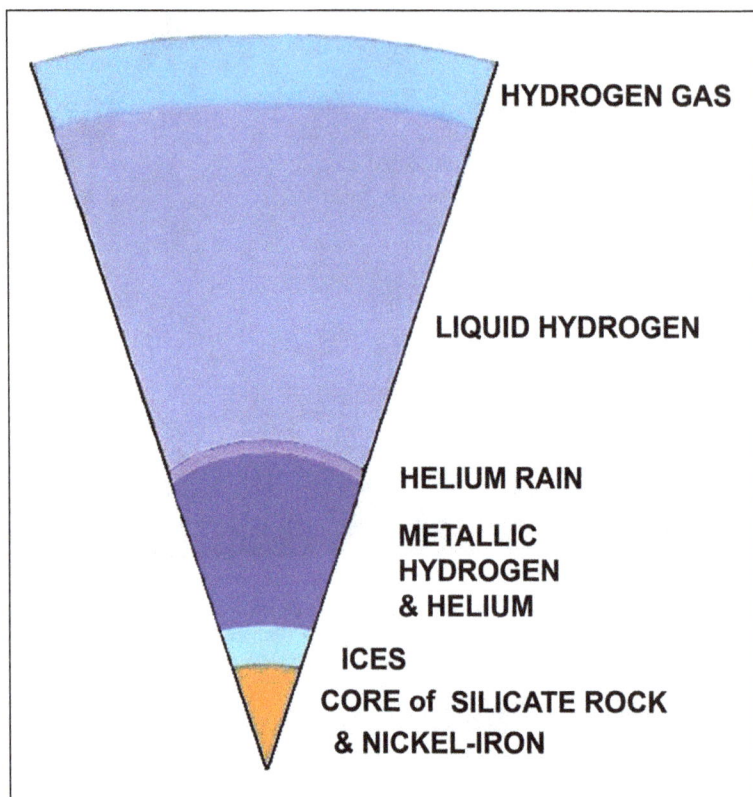

Figure 5.41: Diagram showing a probable internal structure of Saturn

# 5.10 Uranus (Symbol: ⛢ Representing the Power of the Sun and the Spear of Mars)

Figure 5.42: Uranus seen from *Voyager* 2 in visible light as a featureless planet (Photo: NASA)

Figure 5.43: Uranus seen in Infra-red using the Hubble Space Telescope which clearly shows its rings and axial tilt (Photo: NASA)

Discovered in 1781 by **Sir William Herschel** (German/British: 1738 -1822), Uranus is named after the Greek god of the sky, Ouranos. It has a different chemical composition from that of the larger gas giants Jupiter and Saturn, and is classed as an **ice giant** as it is composed of greater percentages of ices, mainly of water, methane and ammonia, in its atmosphere. The rest of its composition consists of hydrogen and helium similar to that of Jupiter and Saturn. Uranus has the coldest atmosphere in the Solar System, with a minimum temperature of 49 K (–224.2 °C), and a complex, layered cloud structure. Uranus also has a faint ring system, a magnetic field and numerous moons. The axis of rotation of Uranus is tilted sideways, nearly into the plane of its orbit about the Sun and so its north and south poles lie on their sides and the planet appears to roll along its orbital path. Uranus has 27 known moons, all named after characters from the works of Shakespeare and Alexander Pope. The five main satellites are Miranda, Ariel, Umbriel, Titania, and Oberon.

Some of the main facts about Uranus are:

- average distance from the Sun of 2869.6 million kilometres (19.18 A.U)

- orbital period (year) of 84.07years

- average orbital velocity of 6.80 km/sec

- sidereal rotation period (day) of 17.9 Earth hours

- equatorial rotation velocity of 9,320 km/hour

- equatorial diameter of 51,118 km

- axis of rotation is tilted at 97.77° to the ecliptic

- surface area of $8.1156 \times 10^9$ km$^2$ (15.91 Earth's)

- volume of $6.833 \times 10^{13}$ km$^3$ ( 63.086 Earth's)

- mass of $8.6810 \times 10^{25}$ kg (14.536 Earth's)

- average density of 1.27 g/cm$^3$

- gravitational acceleration 8.69 m/s$^2$

- escape velocity of 21.3km/sec

- active tectonism not observed due to deep cloud cover

- magnetic field is tilted 59° away from the planet's spin axis Uranus and it varies from place to place –in its southern hemisphere, the magnetic field is only about one third that of Earth's field but in parts of the northern hemisphere it is almost four times as strong as Earth's

- surface temperatures is about 57K (-216.15 °C) to 76 k (–197.2 °C)

- atmospheric pressure in the upper cloud layer is very high (> 1000 bars) with a composition of:

|               |        |
|---------------|--------|
| hydrogen      | 83.0%  |
| helium        | 15.0%  |
| methane ($CH_4$) | 2.3 % |

$$\text{ammonia (NH}_4) \qquad 0.01\%$$

with smaller traces of hydrogen deuteride and ices of ammonia, water, ammonium hydrosulfide ($NH_4SH$) and methane hydrate.

- natural satellites of 27 moons with the largest, Titania, only having a radius of only 789 km, there is also a faint series of rings of dust and ice.

Figure 5.44: False-coloured image of Oberon, Uranus' largest moon taken by *Voyager 2* in 1986 (Photo: NASA)

The density of the planet of 1.27 g/cm$^3$, suggests that its interior is made up of mainly various ices, such as water, ammonia, and methane. Hydrogen and helium constitute only a small part of the total with the remainder of the non-ice mass being silicate and iron-nickel rock. Its structure is probably of three layers: a small *silicate/iron-nickel* rock core with a radius of less than 20% of the planet, covered by a large mantle consisting mainly of a hot, dense fluid of water and ammonia with high electrical conductivity. Above the mantle is a small outer gaseous layer of hydrogen and helium (also about 20% of the radius). Internal pressure is about 8 million bars in the centre which has a temperature of about 5000 K. It has been suggested that the pressures within the atmosphere and electrical discharge may break up the methane molecules ($CH_4$) into crystals of diamond (C) which then fall down to the surface where they float like solid 'diamond-bergs'.

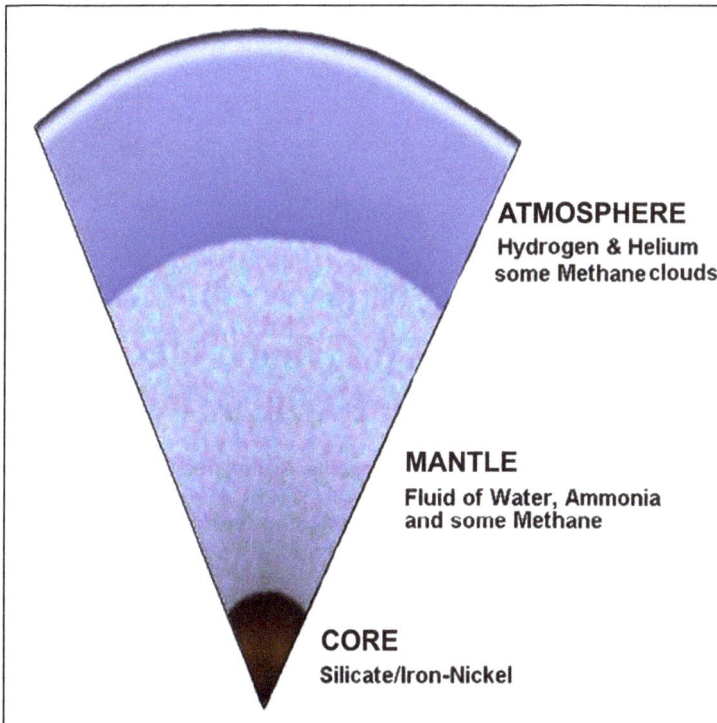

**ATMOSPHERE**
Hydrogen & Helium
some Methane clouds

**MANTLE**
Fluid of Water, Ammonia
and some Methane

**CORE**
Silicate/Iron-Nickel

Figure 5.45: Diagram showing a model for the interior of Uranus

## 5.11 Neptune (Symbol: ♆ The Trident of the Roman God of the Sea)

Figure 5.46: Neptune showing the Giant Dark Spot storm in its upper atmosphere. This photo was taken by *Voyager 2* in 1989. (Photo: NASA).

Neptune is the furthest and eighth major planet from the Sun at a distance of 4,497.1 million km and is not visible to the naked eye. It is named after the Roman god of the sea and was its position was predicted by **Urbain Le Verrier** (French: 1811 –1877) in 1846 following calculations of fluctuations in Uranus' orbit suggesting the presence of another, large body nearby. The planet was finally observed through a telescope later that year by **Johann Galle**

(German: 1812 – 1910). In contrast to the hazy, featureless atmosphere of Uranus, Neptune's atmosphere has active surface cloud patterns such as the **Great Dark Spot** in the southern hemisphere seen during the 1989 *Voyager 2* flyby. Neptune also has a faint and fragmented ring system of arcs which was detected by *Voyager 2*. Amongst the bigger, outer planets of the Solar System, Neptune is the densest.

Some of the main facts about Neptune are:

- average distance from the Sun of 4,497.1 million kilometres (30.06 A.U)

- orbital period (year) of 164.81years

- average orbital velocity of 5.43 km/sec

- sidereal rotation period (day) of 19.1Earth hours

- equatorial rotation velocity of 9660 km/hour

- equatorial diameter of 48,600 km

- axis of rotation is tilted at 28.32° to the ecliptic

- surface area of $7.6183 \times 10^9$ km$^2$ (14.98 Earths)

- volume of $6.254 \times 10^{13}$ km$^3$ (57.74 Earths)

- mass of $8.6810 \times 10^{25}$ kg (14.536 Earths)

- average density of 1.638 g/cm$^3$

- gravitational acceleration 11.15 m/s$^2$

- escape velocity of 23.5 km/s

- active tectonism not observed due to deep cloud cover

- magnetic field is relatively large and is produced by electric currents within the planet. It is tilted at 47° to the planet's rotation axis and also offset by 13,500 km from the planet's centre. It is probably relatively closer to the surface and so Neptune's magnetic field goes through dramatic changes as the planet rotates in the solar wind.

- surface temperatures is about 52k (-221.2 °C) to 75 k (–198.2 °C)

- atmospheric pressure in the upper cloud layer is very high (> 1000 bars) with a composition of:
  hydrogen          80.0%
  helium            19.0%
  methane (ch$_4$)      1.5 %
  ammonia (nh$_4$)      0.01%

  with smaller traces of hydrogen deuteride and ethane (C$_2$H$_6$) and ices of ammonia, water, ammonium hydrosulfide (NH$_4$SH) and possibly methane.

- Natural satellites consist of 14 moons with the largest, Triton (named after the Greek god and son of Neptune) having 99.5% of the total mass of all moons, has a very faint atmosphere and a retrograde orbit. This may suggest that it was captured by the planet from the

Kuiper belt. There is also a faint system of rings consisting of dust and ice.

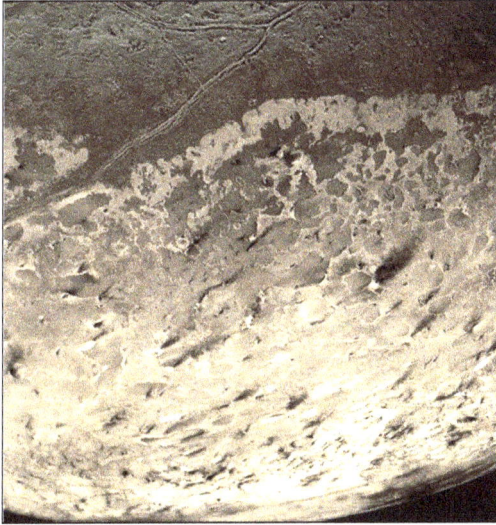

Figure 5.47: Enhanced photograph of Triton's south polar region showing some dark plumes (lower right) which may be dust deposits from nitrogen geysers (Photo: NASA from *Voyager 2* flyby).

Neptune is similar in composition to Uranus, with Neptune's atmosphere mainly composed of hydrogen and helium along with traces of methane and possibly nitrogen. It also contains a high proportion of the ices of water, ammonia, and methane which gives a bluish appearance to the planet. It is believed to have a rocky core. Because of its great distance from the Sun, Neptune's outer atmosphere is one of the coldest places in the Solar System, with temperatures at its cloud tops approaching 55 K (−218 °C). Temperatures at the planet's centre are approximately 5,400 K (5,100 °C).

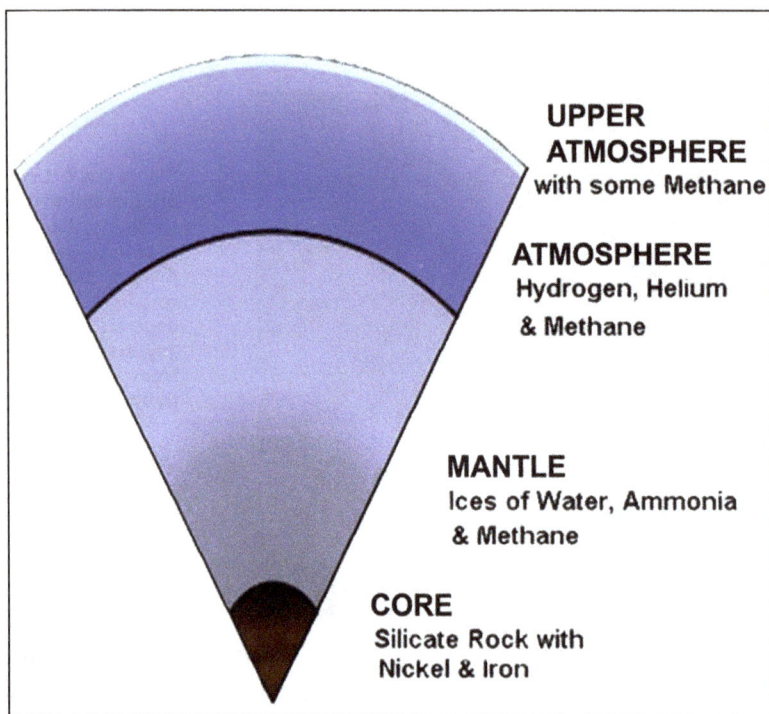

Figure 5.48: Diagram showing the possible interior of Neptune

## 5.12 The Minor Planets

The term Minor Planet has been used in a vague and sometimes misleading sense for any astronomical body which orbits the Sun but is not a comet. Since 2006, the International Astronomical Union (IAU), which acts as the authority in identification and naming of astronomical objects, has used the following as official terms:

- dwarf planets

- asteroids

- Trojans

- centaurs

**Dwarf planets** are those small planets which orbits the Sun, are massive enough for its shape to be in equilibrium under its own gravity and thus nearly rounded in shape, but still shares its orbit with a considerable amount of debris. This is unlike the major planets which have more mass in themselves than exists within their orbits.

The International Astronomical Union (IAU) currently recognizes five dwarf planets:

- Ceres is the largest object in the main asteroid belt, which lies between the orbits of Mars and Jupiter and has a diameter of about 945 km. Following the predictions of the Titus-Bode law in 1772, this small planet was discovered by **Giuseppe Piazzi** (Italian: 1746-1826) in 1801 and named after the Roman goddess of agriculture. It has an orbital period of 4.6 years and a rotational period (day) of 467 days and a mass of $9.393 \times 10^{20}$ kg.

Figure 5.49: The dwarf planet Ceres seen from the
*Dawn* spacecraft in 2015 (Photo: NASA)

- Pluto is in the Kuiper Belt, beyond the planet Neptune, and its orbit is highly elongated, ranging from 30 to 49 A.U. and its orbit is greatly inclined at $17^{\circ}$ to the ecliptic. Sometimes its elongated and inclined orbit puts Pluto within the orbit of Neptune. Pluto was discovered by **Clyde Tombaugh** (American: 1906 –1997) in 1930 from consecutive movement of the image seen on photographic plates. It was originally considered as the ninth planet from the Sun, but after 1992 its status as a planet fell into question following the discovery of several objects of similar size in the Kuiper belt. It has a diameter of 2,370 kilometers, an orbital period of 248 years, a rotational period of 367 days, and a mass of $1.303 \times 10^{22}$ kg. Pluto has a trace atmosphere of nitrogen ($N_2$), methane ($CH_4$), and carbon monoxide (CO) with a surface pressure of from 0.65 to 2.4 Pa (compared to Earth's 101325 Pa). It is believed that Pluto's internal structure consists of a dense rocky core about 70% of its diameter which is surrounded by a mantle of water ice. Pluto has five known moons: Charon (identified in 1978, spherical with a mass about 12% of its planet), Nix (2005), Hydra (2005), Kerberos (2011) and Styx (2012). The four smaller moons are all irregular in shape.

- Haumea was discovered in 2004 and 2005 by two independent teams at Caltech (USA) and the Sierra Nevada Observatory (Spain) and was named after the Hawaiian goddess of childbirth. It has a slightly, elongated shape possibly due to a collision in the early days of the Solar System, an eccentric orbit of from 34 A.U. to 51 A.U. and an orbital period of 284 Earth years. It has a mass of $4.0 \times 10^{21}$ kg and two very small moons, Hiʻiaka and Namaka.

Figure 5.50: Pluto and its largest moon, Charon in relative size (montage from NASA images)

- Makemake was discovered in 2005 by a team at Mount Palomar Observatory (USA) and is one of the larger

Kuiper belt objects with a diameter that is about two-thirds the size of Pluto. Its name derives from the god of fertility of the Rapa Nui people of Easter Island. It has an elongated orbit (38-52 A.U.) and an orbital period of 309 years. It has no known satellites, and a covering of methane, ethane, and possibly nitrogen.

- Eris was also discovered in 2005 by the Mount Palomar team and is the most massive and second-largest of the dwarf planets with a mass 27% more than that of Pluto and a diameter of 2,326 ± 12 km. It is named after the Greek goddess of strife, has a distance from the Sun varying from 37.9 A.U. to 97.6 A.U. and has an orbital period of 558 years. Due to Eris' distant eccentric orbit, its surface temperature is estimated to vary between about 30 K and 56 K (–243.2 and –217.2 °C).

There are other Dwarf planets outside of the orbit of Neptune, along with the last four mentioned above and are collectively called the Trans-Neptunian Objects, but there is still doubt as to their classification as dwarf planets:

- Quaoar (pronounced kwarwar) is named after a Native American creator god, was discovered in 2002, and is about half the size of Pluto. It also has a small moon called Weywot. Its distance from the Sun varies from 41 A.U. to 45 A.U. and it has an orbital period of about 256 years.

- Sedna, named after the Inuit goddess of the sea, was discovered in 2003, and has a very eccentric orbit, varying from 79 A.U. to greater than 936 A.U. and an orbital period of about 11,400 years. It is thought to be smaller than Charon, the moon of Pluto.

- Orcus, discovered in 2004 is named after the Etruscan god of the underworld. It has a distance from the Sun from 30 A.U. to 48 A.U. and an orbital Period of 245 years. It is very bright and may be covered in ice and has a diameter about 800 km. Orcus has one small moon called Vanth.

There are many other minor planets beyond Neptune's orbit, some have a relationship with that planet's gravity, others with that of Pluto and still more are well beyond these planets' orbits. Most are within the Kuiper Belt orbiting the Sun in the ecliptic Plane at distances of from 30 A.U. to 50 A.U.

**Asteroids** are small to large, irregular Minor Planets which orbit the Sun in a variety of orbits. There are millions of asteroids within the Solar System and it is thought that some of these may be the shattered remnants of the early smaller planetesimals within the young Sun's solar nebula that never grew large enough to become large planets. Most orbit in the main asteroid belt between the orbits of Mars and Jupiter, or co-orbit with Jupiter as the Jupiter Trojans and Greeks. There are also smaller groups including the near-Earth asteroids. Individual asteroids are usually given numbers but some are named and all are classified by their compositions as being carbon-rich, stony or metallic. Only one asteroid, Vesta, which has a relatively reflective surface, is visible to the naked eye under extremely favourable conditions.

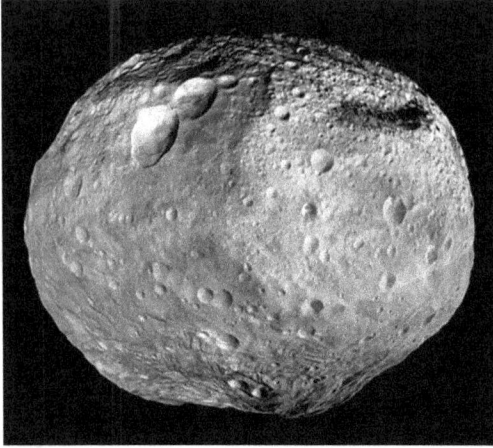

Figure 5.51: Asteroid Vesta photographed by the Dawn Spacecraft (Photo: NASA)

**Trojans** are minor planets or moons which share the orbit of a larger planet such as Jupiter, remaining near one of the two points of stability or Lagrange points which are designated $L_4$ and $L_5$ – situated approximately 60° ahead of and behind the larger body. The **Jupiter Trojans** are the most well-known of the Trojans in the Solar System and they are divided into the Greek camp ($L_4$) in front of Jupiter and the Trojan camp ($L_5$) trailing behind Jupiter in their orbit (see Figure 5.30). There are also a smaller number of Trojans associated with Mars, Neptune and one was discovered within Earth's orbit named 2010 TK7 by NASA in 2011.

**Centaurs** are minor planets which have unstable orbits and lie outside the orbit of Jupiter but well within the Kuiper Belt. Centaurs seem to have some of the characteristics of both asteroids and comets and are named after the mythological Centaurs of Greek legend. The largest confirmed centaur is 10199 Chariklo, which has an estimated diameter of about 260 km and has a system of rings and was discovered in 1997. It is also thought that Saturn's moon Phoebe may be a captured centaur. At least three Centaurs have been found to have comet-like comas and it may be expected that any centaur that is able to come close enough to the Sun is expected to become a comet.

## 5.13 Meteors and Meteorites

**Meteors** named from Ancient Greek *meteōros*, meaning high in the air, are commonly called shooting stars. **Bolides** are larger meteors which can be seen during the day. When these **meteroids**, which vary from dust to asteroid size bodies, asteroids or even comets pass into the Earth's atmosphere, they are rapidly heated by friction with the air and are seen flashing across the sky. Most are usually dust size and are often seen the night sky on most nights as individual tracts of light, but they also can come in predictable showers or groups of many meteors when the Earth passes through a specific patch of space debris or the tail of a comet. These showers are given the name of the star constellation in the night sky in which they first appear. Meteors typically burn up in the atmosphere at altitudes from 76 to 100 km. but occasionally, larger fragments survive the disintegration passing through the Earth's atmosphere and strike the Earth's surface. When they do this, they are termed **meteorites,** and several large craters which have survived Earth's tectonism, sedimentation and erosion can still be seen.

Meteorites have been commonly classified as either:

- Stony meteorites of silicate composition which may be **chondrites** which have not been modified due to melting or differentiation and contain chondrules or round grains or droplets of minerals, or **achondrites** which do not contain chondrules.

Nickel-iron **meteorites** or a composite of both but more modern classifications often consider more detailed chemistry.

Figure 5.52: Nickel-iron meteorite found in Texas, USA, showing the crisscross lines of metallic crystal structure (widmanstätten pattern). (Photo: NASA)

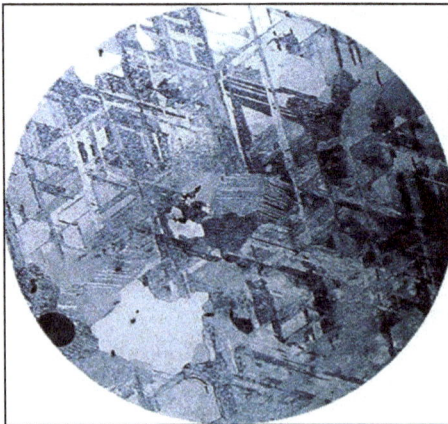

Figure 5.53: magnified view of the widmanstätten pattern (x 10) on a cut and polished section of a meteorite. (Photo: NASA)

Figures 5.54: Perseid meteor shower coming from the Constellation of Perseus (Photo: NASA/JPL)

Figures 5.55 The Barringer meteorite crater in Arizona, USA is 1,200 metres in diameter and some 170 m deep (Photo: USGS/D. Roddy).

## 5.14 Comets

Comets are small icy, irregularly-shaped bodies, made up mainly of ices such as water, carbon dioxide, carbon monoxide, methane and ammonia, and possibly some minor organic compounds such as ethane, formaldehyde, and other hydrocarbons. They orbit the Sun in very elliptical orbits and their name comes from the Greek for wearing long hair. When passing close to the Sun, the increased temperature causes their outer layer to vaporize and this vapour is swept out by the solar wind as a bright tail. The remaining head consists of a bright, diffuse **coma** surrounding the more solid, inner nucleus. A comet's nucleus could range from a few hundred metres to tens of kilometres across and is composed of loose collections of ice, dust, and small rocky particles. The coma and tail are much larger and may be seen from the Earth by the naked eye. **Comets** have a wide range of orbital periods, ranging from several years to potentially several millions of years. Short-period comets originate in the Kuiper Belt but some long- period comets are thought to originate in the distant Oort Cloud, far outside the Kuiper Belt and halfway to the next nearest star. Many of the shorter period comets are predictable and it was **Edmund Halley** (English: 1656 -1742) who in 1705, first predicted the return of the comet which is now named after him. Halley's Comet returns every 75-76 years with the next sighting due in 2061.

Figures 5.56: Halley's Comet photographed in 1986 (Photo: NASA)

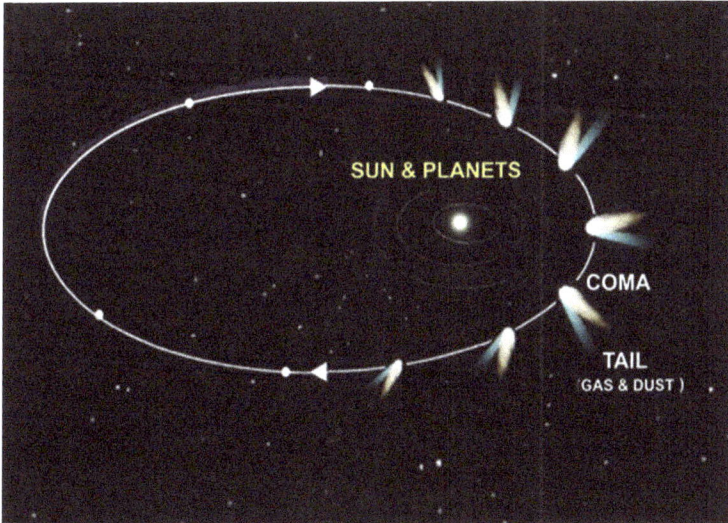

Figures 5.57 A diagram showing the path of a comet around the Sun

Many free Apps. are available (Search Words: **Solar System, Planets, Solar Eclipse, Sun, Telescopes, Comets, Asteroids** – search for Offline Apps. There are some good Apps. which can be directly used with some telescopes for locating sky objects)

# Chapter 6: Beyond to the Stars

## 6.1 Introduction – Nature and Grouping

A star, such as our Sun, is a luminous sphere of plasma held together by its own strong gravitational pull against its expansion by heat and radiating energy in the forms of light, heat, ultra-violet and other types of radiation. Planets, moons and most other objects seen in the night sky are illuminated by the Sun's light which is reflected off them. Other stars which are visible from Earth during the night appear as a multitude of fixed luminous points in the sky of different brightness and colours due to their immense distance from Earth, their size, temperature and composition. Many of the stars seen at night are in fact not single stars but double or triple systems rotating around their centre of mass. Some are **nebulae**, huge luminous gas clouds in which stars are formed whilst others are distant galaxies or **star clusters**.

Observation and curiosity about the stars goes back before recorded history. Civilizations have known that the stars seem to hold fixed positions in the sky, some as recognized patterns which could be identified with religious or mythological figures useful in rituals or as an aid to navigation. These constellations, from the Latin *constellātiō* for stars together, were believed to be fixed upon a celestial sphere which rotated over the Earth from east to west. Constellations were also used as reference points to track the path of the Sun and planets and develop the concept of time and the use of calendars. The Gregorian calendar currently used nearly everywhere in the world, is based on the angle of the Earth's rotational axis relative to its local star, the Sun. These night observations and recordings were useful in making valid predictions about seasonal change for crop planting and animal migration.

Many of the ancient astronomers, and several after the renaissance, such as **Tycho Brahe** (Danish: 1546-1601), **Nicholas Copernicus** (Polish: 1473-1543) and **Johannes Kepler** (German: 1571-1630) still followed the older traditions of the astrologers in attempting to use the position of the constellations, Sun, Moon and planets as aids in making predictions using their charts or horoscopes, about future social and personal events. Whilst modern science does not give any credibility to astrology, the contribution made by early astrologers to modern astronomy in their observations, detailed measurements and ideas, has been incredibly valuable.

## 6.2 Locating Stars

Mapping stars is as important as mapping geological and geographical features because there is a need to record the data and to communicate it to other interested parties. Just as in using geographical coordinates (latitude and longitude) to map the surface of the Earth, the sky also has a coordinate system. In astronomy, the celestial coordinate system most favoured by astronomers is the **equatorial coordinate system**, which is used to align telescope mountings, and is the projection of geographical latitude and longitude onto the **celestial sphere**. Lines of latitude are projected as lines of **declination** (Dec), measured in degrees and minutes and seconds of arc, which indicate how far north or south the sky object is from the celestial equator which is defined by projecting the Earth's equator onto the celestial sphere. Lines of longitude are projected as lines of **right ascension (RA),** and are measured in hours, minutes and seconds from where the celestial equator intersects the ecliptic. This intersection is called the **vernal equinox** (or the First Point of Aries or Spring Equinox) and is the celestial prime meridian from which right ascensions are measured (i.e. is zero hours RA). Exactly at 180° (or 12 hours RA) from it is the First Point of Libra.

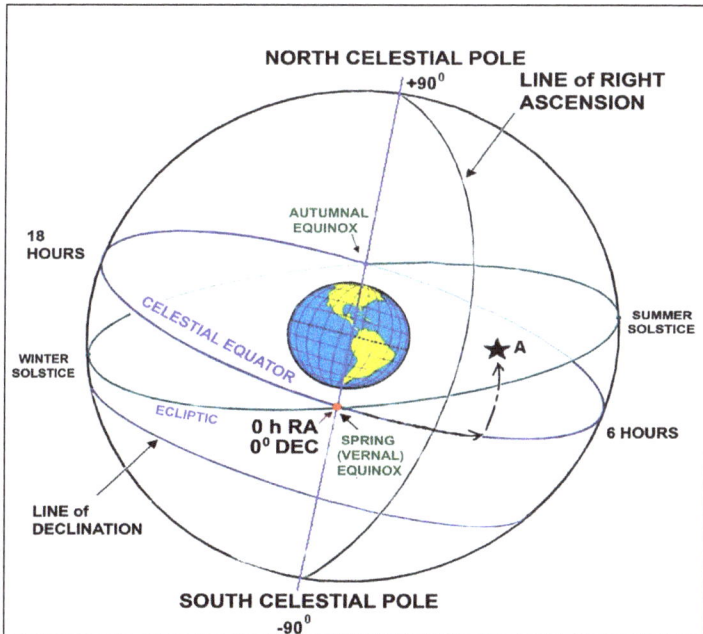

Figure 6.1: Diagram showing the Ecliptic Coordinate System for locating stars. The star at A would have a location of about 3 hours RA and +30° declination.

Another way of communicating locations of objects in the night sky is by using the **horizontal coordinate system**. This is a more casual way of communicating information between observers who are at about the <u>same location</u> and observing at about the <u>same time</u>. This system uses **azimuth**, the geographical bearing $000^0$ to $360^0$ with North at these values, East at $090^0$, and South at $180^0$ and so on, and **altitude**, the angular height of the object above the horizon from $00^0$ at the horizontal to $90^0$ at vertical. An observer would have to give the time and location as well as the azimuth and altitude to indicate the position of the object.

## 6.3 Constellations

Constellations are recognizable patterns of stars seen in the night sky. The International Union of Astronomers lists 88 constellations seen around the world today. The earliest evidence for constellations and star catalogues comes from inscribed stones and clay tablets found in Mesopotamia (within modern Iraq) dating back to 3000 BC. And the oldest accurately dated star chart appeared in ancient Egyptian astronomy around 1534 BC.

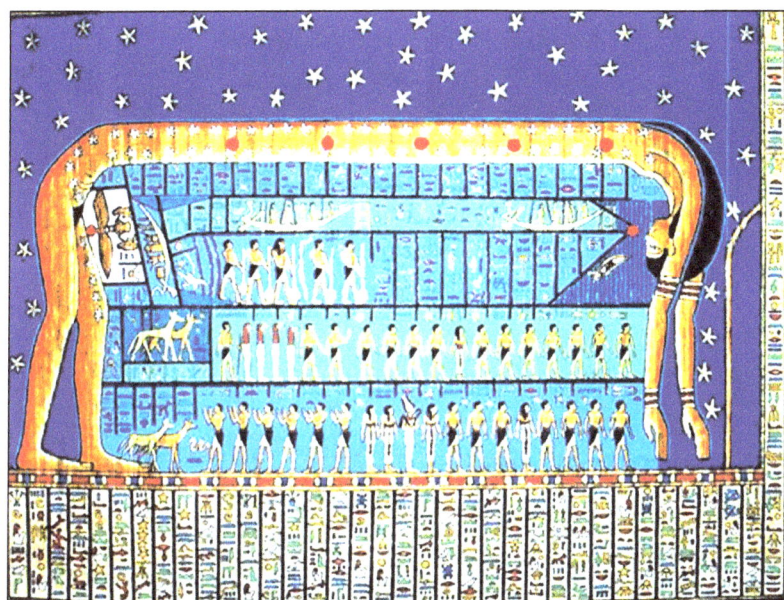

Figure 6.2: Nut, the Egyptian goddess of the sky protects all below
(Photo:Wiki Commons)

Of great importance to the early astrologers, where the constellations of the zodiac, named from the Greek *zodiakos kyklos* meaning circle of little animals, which moved on a pathway through the stars which the Sun and the planets appear to take due to the rotation of the Earth. The 12 signs of the **zodiac** linked to their constellations which move across the sky along the ecliptic in order are:

**Aquarius (The Water Bearer)**
**Pisces (The Fish)**
**Aries (The Ram)**
**Taurus (The Bull)**
**Gemini (The Twins)**
**Cancer (The Crab)**
**Leo (The Lion)**
**Virgo (The Maiden)**
**Libra (The Scales)**
**Scorpio (The Scorpio)**
**Sagittarius (The Centaur Archer)** and
**Capricorn (The Sea-Goat).**

These twelve signs of the zodiac were used by astrologers for their non-scientific calculations. Astronomers also use them for star references, and since 1930, an additional zodiac constellation, Ophiuchus (The Serpent-bearer) has been added between Scorpio and Sagittarius.

Figure 6.3: A chart of the Zodiac on the ceiling of a roof-top room on the Temple of Hathor at Dendera, Egypt, dating from about 30 B.C.

Many of these were named after ancient legends in societies where knowledge was passed on verbally rather than being written down and so were used as a memory aid. For example, the Greek legend of Orion the Hunter tells of a hunter who boasted that he was invincible and that no animal could harm him. The Greek gods were displeased with this boast and sent a scorpion (Scorpio), the only animal which could harm Orion, to sting him. When they saw Orion die, they had pity upon him and so, as was the common with the Greek gods, placed both Orion and the scorpion in the sky so that they would be forever apart. In the summer skies of the southern hemisphere, the constellation of Orion tracks westward across the sky followed by the constellation of Scorpio which only rises as Orion sets. Other ancient peoples had different myths to tell their children, often with some moral or more practical reason. For example, the Yolngu people of northern Australia say that the constellation of Orion (Djulpan) is a canoe used by three brothers who went fishing. One of them ate a fish which was forbidden under their law and seeing this, the sun-woman (Walu), made a waterspout that carried the three and their canoe up into the sky. Scientifically, the constellations are good as references for the locating main stars, planets and other celestial objects.

Much of our modern practices and references to stars date from the studies of the Ancient Greeks who also had used the earlier writings of the Mesopotamians in their astronomy. The first Greek star catalogue was attributed to **Aristillus** (about 300 B.C.) and by the 2nd century BC, the star catalogue of **Hipparchus** (Greek: 190- 120 B.C.) included over one thousand stars. Hipparchus is also known for the first recorded sighting of a **nova** (From Latin for new and specifically referring to a new star) – a sudden brightening of a star. From these ancient Greek observations as well as his own, **Claudius Ptolemy** (Greco-Egyptian: 100 – 170 A.D.) compiled his great catalogue and collection of observations and ideas called *The Almagest*, which became the major reference for Renaissance astronomers as many of the works of the earlier Greeks had been lost.

| LETTER | SYMBOL | PRONOUNCED | LETTER | SYMBOL | PRONOUNCED |
|--------|--------|------------|--------|--------|------------|
| alpha | α | AL-phuh | nu | ν | noo |
| beta | β | BEE-tuh | xi | ξ | zeye |
| gamma | γ | GAMM-muh | omicron | o | Ohm-MY-kronn |
| delta | δ | DELL-tuh | pi | π | pie |
| epsilon | ε | EPP-sih-lonn | rho | ρ | row |
| zeta | ζ | ZEE-tuh | sigma | σ | SIGG-ma |
| eta | η | ee-tuh | tau | τ | tow |
| theta | θ | THEE-tuh | upsilon | υ | UP-Si-lonn |
| iota | ι | eye-OH-tuh | phi | φ | fie, |
| kappa | κ | CAPP-uh | chi | x | ki |
| lambda | λ | LAMM-duh | psi | ψ | psi |
| mu | μ | moo | omega | ω | OM -ega |

Table 6.1: The Greek Alphabet

Medieval Islamic astronomers gave Arabic names to many stars and these are still used today in addition to the system of names invented by the German astronomer **Johann Bayer** (1572 -1625) in 1603. In this systematic approach to naming stars, Bayer suggested the use of the

Greek Alphabet for the stars within a constellation in order of brightness with the brightest being Alpha, the next being Beta and so on. For example, the brightest star in the constellation of Orion, called Betelguese or in Arabic ابط الجوزاء, *Ibt al-Jauzā'*, meaning the hand of Orion, is called Alpha Orionis (α-Orionis or α- Ori).

Once that all of the Greek alphabet have been used in tagging stars in a constellation, astronomers then use numbers as well. For example, 51 Pegasi in the constellation of Pegasus, is similar to our Sun and has at least one planet orbiting around it. This method of designating stars using numbers and their constellation, first appeared in the publication *Historia Coelestis Britannica* by **John Flamsteed** (English: 1646 – 1719) and published by Edmond Halley and Isaac Newton in 1712 without Flamsteed's approval. Some stars are also named after their discoverer, such as Barnard's Star, named after the American Astronomer **Edward Barnard** (1857 – 1923) who first measured its motion in 1916. This system of naming is no longer used.

Many free Apps. are available (Search Words: **Star Charts; Star Charts; Night Sky** – search for Offline Apps.)

## 6.4 Some Common Constellations

The following constellations are the most easily identifiable and contain some very interesting sky objects. Those which can be seen in both the northern and southern hemispheres will have different orientations, with those of the southern hemisphere being upside down and reversed_sideways and to those originally named in the Northern Hemisphere. Diagrams below are for the Southern Hemisphere:

- **Orion the Hunter**

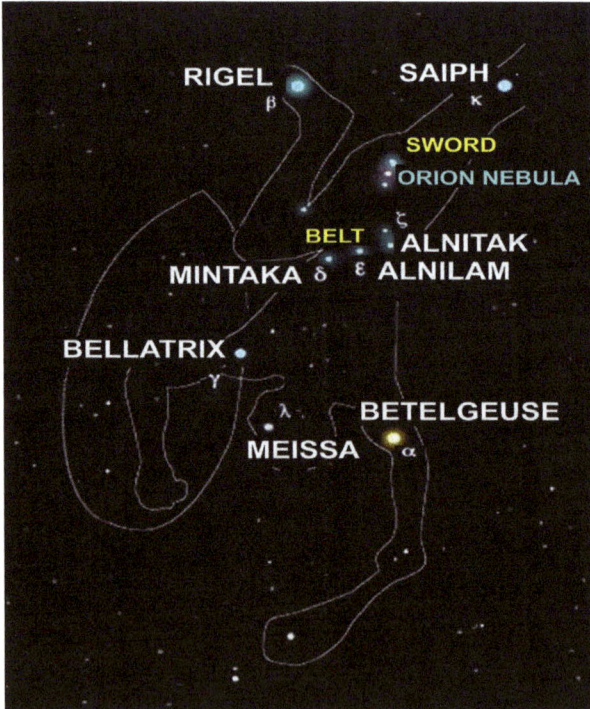

Figure 6.4: Orion the Hunter as seen from the Southern Hemisphere (where its centre is also called the Saucepan). To see the view from the Northern Hemisphere, flip vertical and horizontal.

The Sword Scabbard (also called the Handle of the Saucepan in the southern hemisphere) just apart from Orion's Belt, has three bright stars, the central one being the Great Nebula Of Orion – a large collection of stars and gas which can be seen beautifully with a small telescope.

- **Taurus the Bull**

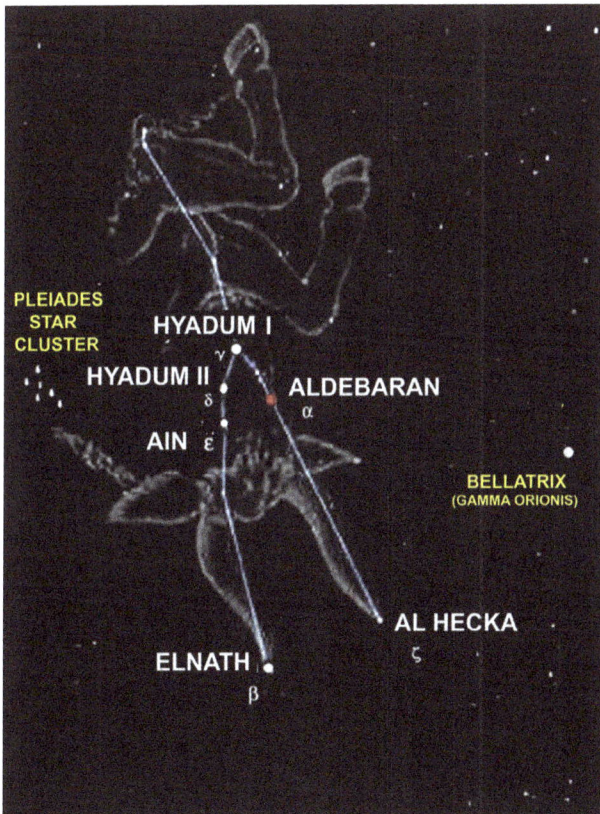

Figure 6.5: Diagram showing the main stars of Taurus as seen in the Southern Hemisphere (flip horizontal & vertical for Northern Hemisphere)

To the west, in front of Orion the Hunter as if to confront him, is Taurus the Bull with its prominent red star, Aldebaran (Alpha Tauri) for its eye. Just past Taurus is the Pleiades, which is an open cluster of seven stars often called the Seven Sisters after the seven nymphs who were the companions of the Greek goddess Artemis.

Figure 6.6: The Pleiades in Taurus, or Seven Sisters named after the seven daughters of the sea-nymph Pleione of Ancient Greek legend (Maia, Electra, Target, Alcyone, Celaeno, Sterope and Merope). Her name comes from the Greek *Plein* meaning sail and probably signifies the beginning of the sailing season in the Mediterranean as these stars first appear in the east (Photo: NASA/ESA/AURA/Caltech)

They are called *Subaru* in Japanese meaning to unite. Taurus is also seen upside down from the Southern Hemisphere.

- Canis Major- the Great Dog

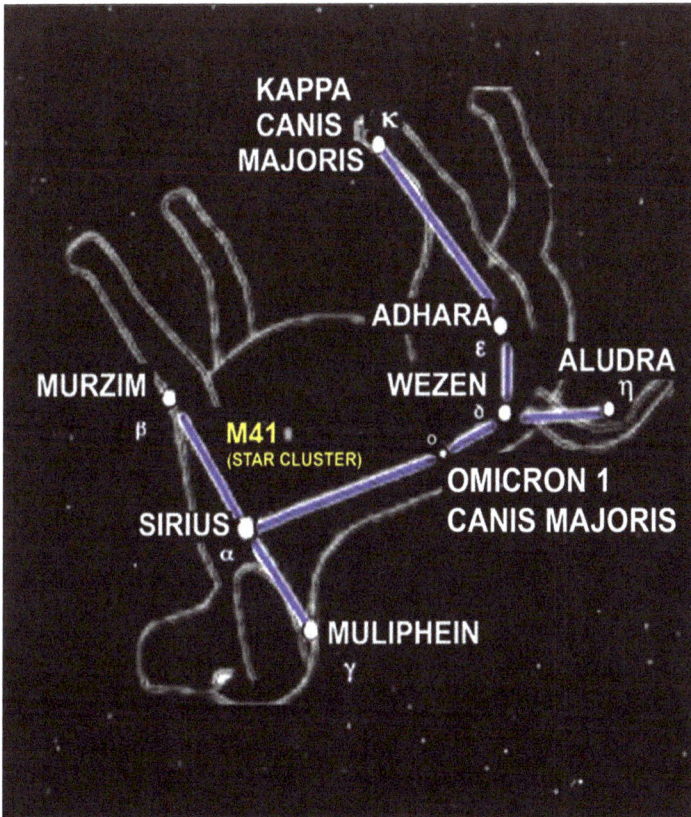

Figure 6.7: Diagram of Canis Major showing the main stars (flip horizontal & vertical for the Northern Hemisphere)

Coming behind Orion the Hunter in the east are his faithful dogs, Canis Major, Latin for The Great Dog and Canis Minor, the Little Dog. Canis Major is one of the few constellations which easily resemble its name, although it looks more like a simple stick figure dog in the night sky which again is upside down in the Southern Hemisphere. Sirius (Alpha Canis Majoris) is named from the Ancient Greek *seirios* meaning glowing and is the brightest star seen in the Southern Hemisphere. It is in fact a **binary star system** of two stars – Sirius A and Sirius B - orbiting each other and is often called The Dog Star. The Dogon people of Mali, Africa were reported by French anthropologists, to have knowledge of the faint Sirius B which would require a telescope to see, but there has been much doubt as to the accuracy of the anthropologists' opinion. Procyon (Alpha Canis Minoris) in nearby Canis Minor is also a very bright star because it is one of the closest neighbours to our solar system at about 11.46 light years away.

- **Scorpio the Scorpion**

Well behind Orion and his dogs, chasing him across the sky but never seen in the sky with him, is Scorpio the scorpion. This constellation is seen later in the night in the southern hemisphere in summer (January) or earlier in winter (June). In the sky one sees it as a giant, inclined hook with the red star Antares or Alpha Scorpii (from the Ancient Greek meaning the equal of Ares, or Mars as it resembles this red planet), near its centre.

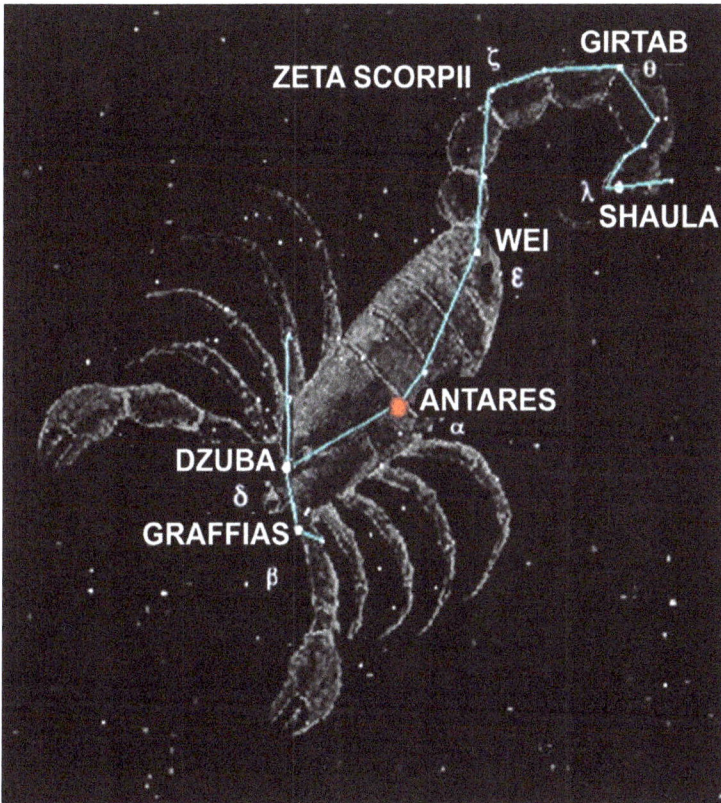

Figure 6.8: Diagram showing Scorpio and its main stars seen from the Southern Hemisphere (flip horizontal & vertical for Northern Hemisphere)

- **Crux, the Southern Cross**

Whilst this is one of the most popular of the Southern Hemisphere constellations and features on the flags of Australia and New Zealand, it was known to the Ancient Greeks who could see it far to the south from

195

Egypt in the summer of the Northern Hemisphere. In the Southern Hemisphere, it is best viewed during winter (June) when it is higher in the sky and is most useful in finding south. This can be achieved because as its long axis rotates around the **South Celestial Pole** which is above the geographical South Pole. Nearby are the two brightest stars of the constellation of the Centaur, Alpha and Beta Centauri, which are also called The Pointers because they are used in finding south using Crux. Alpha Centauri (or Toliman), the brightest of these two stars, is a triple star system comprising Alpha Centauri A and B and the smaller Proxima Centauri which, at 4.24 light years (the distance that light travels in one year at 300,000 km/second), is the closest star to our Solar System.

Geographical South is found from the Southern Cross by extending the long axis of the cross down through its brightest star (Acrux) for four and a half times its length to get to the south celestial pole which is above the geographical South Pole. Crux is a circumpolar constellation, which means that it can easily be seen to rotate around the South Celestial Pole. However, it can sometimes be very low on the southern horizon and difficult to see.

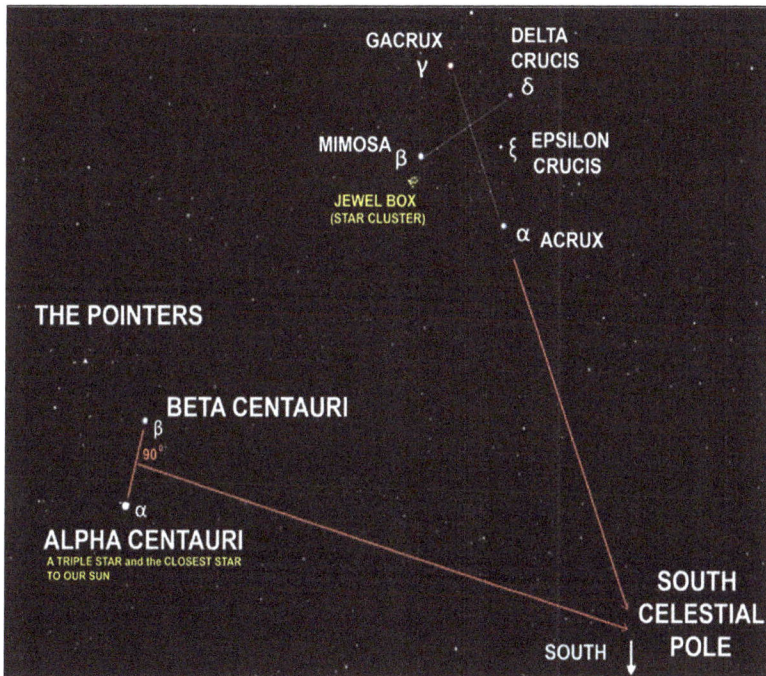

Figure 6.9: Diagram showing The Southern Cross (Crux) and its main stars (flip horizontal & vertical for Northern Hemisphere)

- Leo the Lion

This constellation's name comes from the first of the twelve labours of Hercules and is Latin for lion. In this task, Hercules was required to kill the large Nemean lion. The pattern in the sky forms a long rectangle with a tail reminiscent of a crouching Lion but seen upside down in the Southern Hemisphere. The stars of Leo's mane are an **asterism** or part of a constellation showing a separate well-known pattern; here it is known as the Sickle, which resembles a backwards question mark.

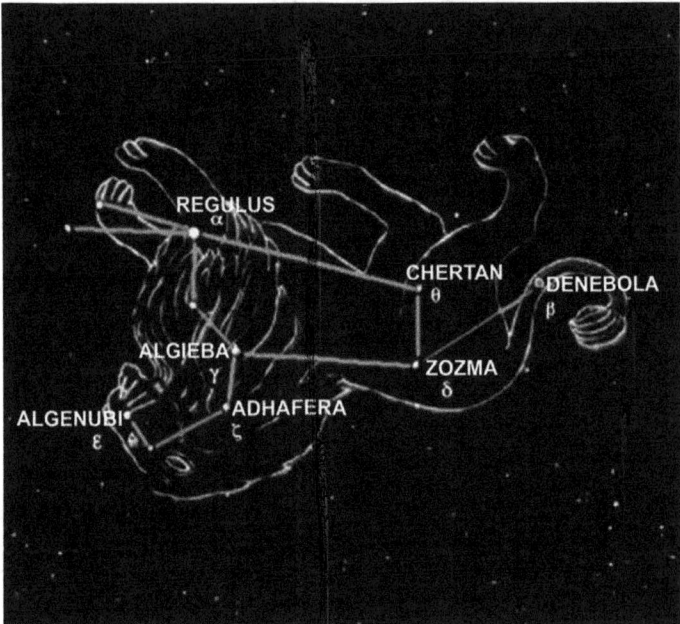

Figure 6.10: Diagram showing the constellation of Leo and its main stars as seen from the Southern Hemisphere (flip horizontal & vertical for the Northern Hemisphere)

- **Ursa Major and Ursa Minor (the Great and Little She Bears)**

These are Northern Hemisphere constellations which have been known since ancient times for their use in navigation. Polaris, (the North Star or Alpha Ursa Minoris) is directly above the geographical North Pole and can be found by firstly locating the bright stars

Dubhe (Alpha Ursa Majoris) and Merak (Beta Ursa Majoris) in the Big Dipper which is another asterism also called The Plough. A line taken from Merak through Dubhe will point to Polaris and directly below this star on the horizon is north.

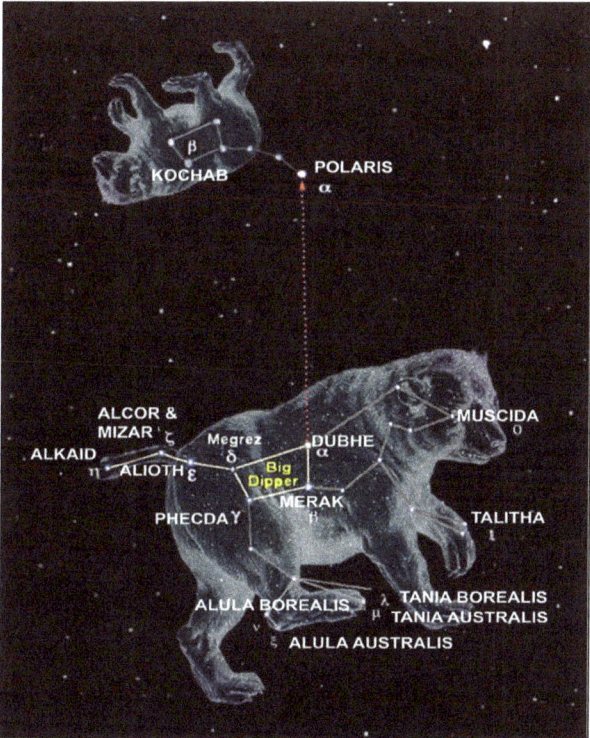

Figure 6.11: The Northern Hemisphere constellations of Ursa Major and Ursa Minor showing the direction to Polaris, the North Star.

## 6.5 Other Constellations

There are about 88 recognizable constellations in the night sky. The most common are:

| NAME | MEANING | BRIGHTEST STAR |
|---|---|---|
| Andromeda | Andromeda | Alpheratz |
| Antlia | Air Pump | α Antliae |
| Aquarius | Water Bearer | Sadamelik |
| Aquila | Eagle | Altair |
| Aries | Ram | Hamal |
| Auriga | Charioteer | Capella |
| Boötes | Herdsman | Arcturus |
| Cancer | Crab | Altarf |
| Capricornus | Sea-goat | Deneb Algedi |
| Carina | Keel | Canopus |
| Centaurus | Centaur | Toliman |
| Cygnus | Swan | Deneb |
| Eridanus | River | Achernar |
| Gemini | Twins | Pollux |
| Hercules | Hercules | Rasalgethi |
| Leo | Lion | Regulus |
| Libra | Scales | Zjbenelgenubi |
| Lyra | Lyre | Vega |
| Microscopium | Microscope | γ Microscopii |
| Pegasus | Pegases | Enif |
| Piscis Austrinus | Southern Fish | Fomalhaut |
| Sagittarius | Archer | Kaus Australis |
| Virgo | Young Girl | Spica |

Table 6.2: Some other important constellations and their brightest stars

## 6.6 Brightness of Stars

Away from light and dust pollution, stars can be seen as having a great variety in brightness. Sirius (Alpha Canis Major), Canopus (Alpha Carinae in Carina, the keel) and Toliman (Alpha Centauri) are three of the brightest stars in the sky. On the other hand, a search of the darkest part of the sky with a small telescope will reveal many faint stars which cannot be seen with the naked eye. **Hipparchus of Nicaea** (Greek: 190 – 120 BC) was the first to attempt the classification of stars by their brightness as they appeared in the night sky. He grouped stars into six categories or apparent magnitudes (also called Visual Magnitudes) depending on how bright they appeared. The brightest First Magnitude stars were twice as bright as the next brightest stars, Second Magnitude, and so on to the faintest stars of the Sixth Magnitude. With the invention of the telescope, however, astronomers soon realized that there were stars much fainter than the Sixth Magnitude and so a more sophisticated system was needed.

In 1856, **Norman Pogson** (English: 1829 – 1891), using photographic techniques, such as filters to block out unwanted light to measure the amount of light coming from a star, noticed that First Magnitude stars were about 100 times brighter than Sixth Magnitude stars. He suggested that the values of 1 to 6 of Hipparchus be formalized using a logarithmic scale, with every interval of one magnitude equating to a difference in brightness of one-fifth power of ten (i.e. $100^{1/5}$) or roughly 2.512 times. This would mean that a First Magnitude star would have an apparent magnitude about 2.5 times brighter than a Second Magnitude star, $2.5^2$ brighter than a Third Magnitude star, $2.5^3$ brighter than a Fourth Magnitude star and so on. As brighter stars than First Magnitude were found in the southern hemisphere with new European settlement, this new scale soon required values less than 1 (and in fact zero) on the old scale.

The realization that stars are not at a uniform distance from us but are at great and variable distances meant that the concept of **apparent (or relative) magnitude (m)** did not give a true indication of the brightness of stars. For example, Sirius appears to be much brighter than Betelgeuse with apparent magnitudes of -1.46 and +0.5 respectively. However, when distances were able be calculated, they were found to be 8.6 and about 1400 light years distant respectively from our Solar System. That is,
Sirius is brighter because it is also closer to the Earth not because it is absolutely brighter than Betelgeuse!

Now, with better telescopes with graduated mounts, could measure very small angles, and distances to stars could be found by the same technique as that used to find the distance to the Moon - the method of **parallax**. For stars, this uses the radius of the Earth's orbit over a six months period as a baseline (of 1 astronomical unit ) and the measured angle is taken at the telescope mount when looking at the target star which will appear to move against the background of distant stars.

The distance can now be calculated and given in units of **parsecs (pc)**. One parsec is the distance at which one astronomical unit subtends an angle of one arc second and is equal to about 3.26 light-years. For example, the nearest star to our Solar System is Proxima Centauri, which has a parallax of 0.762 arcseconds (seconds of arc) and therefore is 1.31 parsecs away. Knowing the distance to a star and a value for its brightness at the telescope using modern photometric methods, as well as the using the Inverse Square Law for Light (that the light intensity is equal to the reciprocal of its distance squared), a true comparison could be made for the real brightness or **luminosity (L)** at the star's surface. Luminosity is the total amount of energy emitted by an astronomical object per unit time. The S.I. unit of luminosity is measured in joules per second or watts but sometimes can be given as a comparison value to the luminosity of the Sun ($L_\odot$) which is $3.846 \times 10^{26}$ watts (W).

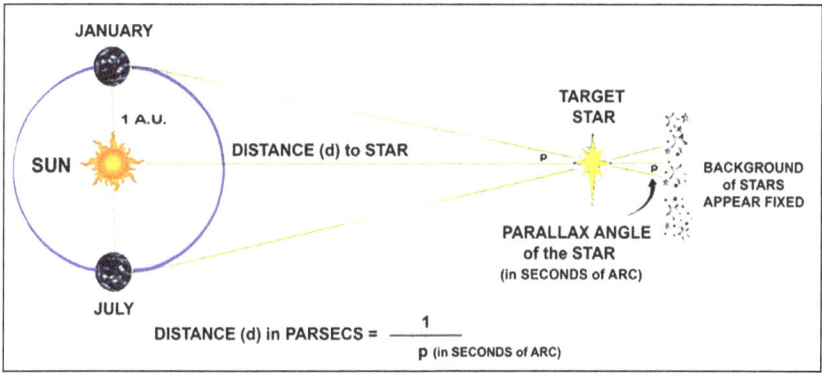

Figure 6.12: Defining the unit of distance called the parsec

As a standard comparison, the luminosity of stars is often given an **absolute magnitude (M)** value which includes a distance factor by converting all values as though the star was at a standard distance of 10 parsecs (i.e. 32.6 light years).

Astronomers can also use the difference between apparent and absolute magnitude (the **distance modulus, μ = *m* - *M***), as a way of calculating distance to a star. If this Modulus is a negative value, the star is closer than 10 parsecs but if it is a positive value it is further away. In general terms, the relationship is:

**Distance Modulus (*m* - *M*) = 5 log(*d*/10)**

> Where **m** = apparent magnitude
> **M** = absolute magnitude and
> **d** = distance in parsecs

For example: Pollux has an apparent magnitude of +1.14 and an absolute magnitude of +0.70 giving a Distance Modulus of (+1.14) – (+1.08) = +0.06. So the equation becomes:

$$0.06 = 5 \log (d/10)$$

$$\text{or } 0.06/5 = \log (d/10)$$

$$\text{i.e. } 0.012 = \log(d/10)$$

but log (d/10) is also log d – log 10 since in logarithms (powers of 10), dividing by logs means <u>subtracting</u> the values (e.g. $10^6 / 10^2 = 10^{6-2}$) i.e. log (d/10) i.e. log (d/10) becomes log d – log 10 and as log 10 = 1, then the equation now becomes:

$$0.012 = \log d - 1$$

$$\text{or } 0.012 + 1 = \log d$$

or **log d** = **1.012**

so from finding the antilog of this value (that power of 10 giving this value):
= **10.3 parsecs**
(or 10.3 x 3.26 Light years)
= **34.5 light years**

Some of the most common magnitude vales for common stars are given in the following table (Table 6.3):

| COMMON NAME | BAYER NAME | APPARENT MAGNITUDE (m)* | ABSOLUTE MAGNITUDE (M) | DISTANCE FROM THE SUN (L.Y.) |
|---|---|---|---|---|
| The Sun | (not given) | −26.74 | +4.8 | 0.000 016 |
| Sirius | α Canis Majoris ("The Big Dog") | −1.46 (triple star) | +1.42 | 8.6 |
| Canopus | α Carinae ("The Ship's Keel") | −0.74 | -5.65 | 310 |
| Toliman | α Centauri ("The Centaur") | -0.27 (0.01 + 1.33*) | +4.34 | 4.4 |
| Arcturus | α Bootes ("The Herdsman") | −0.04 | -0.31 | 36.7 |
| Vega | α Lyrae ("The Lyre") | +0.03 (−0.02 - 0.07) | +0.58 | 25.3 |
| Capella | α Aurgae ("The Little Goat") | +0.08 (0.03 - 0.16) | +0.35 | 42.2 |
| Rigel | β Orionis ("Orion, the Hunter") | +0.13 (0.05 - 0.18) | -8.10 | 860 |
| Procyon | α Canis Minoris ("Little Dog") | +0.34 | +2.66 | 11.5 |
| Achernar | α Eridani ("The River") | +0.46 (0.40 - 0.46) | -1.30 | 140 |

| Betelgeuse | α Orionis | +0.42 (0.2 - 1.2) | -5.59 | 640 |
|---|---|---|---|---|
| Acrux | α Crucis ("The Southern Cross") | +0.76 (1.33 - 1.73) | -4.14 | 320 |
| Aldebaran | α Tauri ("The Bull") | +0.86 (0.75 - 0.95) | -0.30 | 65 |
| Antares | α Scorpii ("The Scorpion") | +0.96 (0.6 - 1.6) | -5.20 | 600 |
| Pollux | β Gemini ("The Twins") | +1.14 | +1.08 | 34 |
| Polaris | α Ursa Minoris ("Little Bear") | +1.98 (1.86 - 2.13) | -3.64 | 430 |
| Saiph | κ Orionis | +2.07 | -4.65 | 722 |
| Wei | ε Scorpii | +2.29 | +0.78 | 65 |
| Acrab | β Scorpii | +2.56 | -3.50 | 404 |
| Proxima Centurii | One of the triple stars with Alpha Centauri | +11.05 variable | +15.49 | 4.24 |

Table 6.3: Table showing some of the magnitude values of common stars.
* Stars with a range of magnitudes are Variable Stars.

## 6.7 Colours and Classification

On a very good night away from light and air pollution, one can see about 4000 stars in the night sky. All of them twinkle because their light is refracted many times by disturbances within the layers of the Earth's atmosphere. Most appear to be white but many have distinct colours such as orange, red, white and blue. The cooler stars, about $3000^0$C, are reddish whilst the hottest stars are blue at about $50,000^0$C.These colours are due to their chemical composition and surface temperatures but are also linked to their size and mass. The colours and their surface temperatures are:

| COLOUR | TEMPERATURE Degrees Kelvin (K) | EXAMPLES |
|---|---|---|
| Red | 2,000 - 3,500 | Proxima Centauri |
| Orange | 3,500 - 4,900 | Epsilon Eridani |
| Yellow | 4,900 - 6,000 | Apha Centauri (Toliman), The Sun |
| White-yellow | 6,000 - 7,500 | Alpha Canis Minoris (Procyon) |
| White | 7,500 - 10,000 | Alpha Lyrae (Vega) |
| Blue-white | 10,000 - 28,000 | Beta persei (Algol) |
| Blue | 28,000 - 50,000 | Zeta Orionis (Alnitak) |

Table 6.4: Colours of stars and their temperatures

Since the discovery of the spectroscope and its attachment to telescopes in the 19th century, Astronomers have observed and recorded the spectral lines of stars. These can be compared with laboratory spectral analysis of the 92 natural chemical elements at different temperatures. At the beginning of the 20<sup>th</sup> century, astronomers using photography and spectrographic techniques were interested in the magnitudes and luminosities of stars. In 1911, the Danish astronomer, **Ejnar Hertzsprung** (1873–1967), plotted the absolute magnitude of stars against their colour as a measure of their temperature. Independently in 1913 the American astronomer **Henry Norris Russell** (1877–1957) plotted spectral class against absolute magnitude. Their graphs both suggested that the relationship between temperature and luminosity of a star was not random but instead appeared to fall into distinct groups. Since that time, these graphs have become known as **Hertzsprung-Russell (Or H-R) diagrams** and they combine the quantities measured by both Hertzsprung and Russell. They have become a useful tool in studying the life cycle of stars of different sizes.

Most stars are classified using the **Harvard spectral classification system** which was created by Annie Jump Cannon (American: 1863-1941) through her examination of spectra from 1918 to 1924. Originally, the scheme used capital letters running alphabetically, but was later reordered to reflect the surface temperatures of stars. In order of decreasing temperature, these types were O, B, A, F, G, K. and M. Three additional categories also in the scheme: R, N, and S types were later realized to represent stars with peculiar heavy-metal abundances. Other types not frequently encountered were also later included, e.g. Q for novae and WR for Wolf-Rayet stars which_show unusual spectra with prominent broad emission lines of highly ionized helium and nitrogen or carbon carbon. The following mnemonic to assist students in remembering the scheme:

**Oh Be A Fine Girl Kiss Me**

In this system, **O** stars are the hottest and **M** stars are the coolest. Traditionally, **O** stars are called blue, **B** blue-white, **A** stars white, **F** stars yellow-white, **G** stars yellow, **K** stars orange, and **M** stars red, but these colours may appear differently to an astronomer depending upon the conditions of observation. In a modern refinement (the **Morgan-Keenan System**), each spectrum letter class has been expanded by a number from **0** to **9** indicating tenths of the range between any two star classes. In addition, luminosity is expressed by the Roman numerals

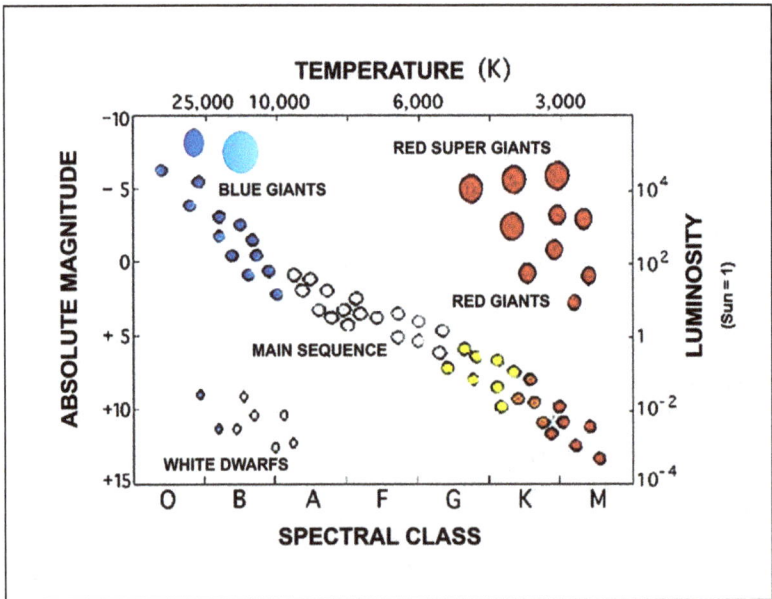

Figure 6.13: H-R diagram showing groups of the most common stars

I, II, III, IV and V, to indicate the width of certain absorption lines in the star's spectrum.

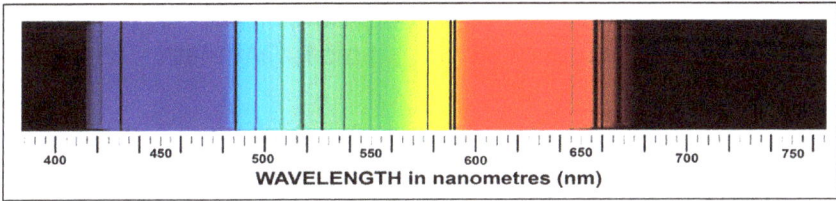

Figure 6.14: Diagram showing an absorption spectrograph for the Sun. The dark lines typical of such spectra are due to wavelengths of light for specific chemical elements which are absorbed by the Sun's atmosphere

The main features of the stars of the different spectral classes are:

- O-Type stars are very hot and extremely luminous, mostly radiating wavelengths in the ultra-violet and blue range. They are the rarest of all main-sequence stars, being only 0.00003% of all the main sequence stars. Many of them are also very massive in size being typically about 60 times the Sun's mass and over ten times its radius. They have high temperatures, usually greater than 25,000 K and very high luminosity (often over a million times that of the Sun). Examples of these stars are 10 Lacertae, and Sigma Orionis of the main sequence and the giants Mintaka (Delta Orionis) and Alnitak (Zeta Orionis).

- B-Type Stars are very luminous and also blue are very energetic (with temperatures from 11,000 to 25,000 K) and live for only a short time. They are often associated with giant interstellar clouds such as in the constellation of Orion and form many of the bright

211

areas of the spiral arms of our Milky Way Galaxy. They make up about 0.125% of main sequence stars and usually have masses about 18 times that of the Sun and luminosities of about 20,000 times that of the Sun. Examples include Rigel, Spica, Alnilam (Epsilon Orionis) and Eta Aurigea.

- A-Type stars are the more common naked eye stars, and are usually white or bluish-white with temperatures ranging from 7,500 to 11,000 K. About three to four times the size of the Sun, they have luminosities of about 80 times that of the Sun. About 0.625% of the main sequence are of this type and include stars such as Sirius, Vega, Deneb (Alpha Cygni – in the constellation of the Swan) and Fomalhaut (Alpha Piscis Austrini – in the constellation of the Southern Fish).

- F-Type stars are white in colour, temperatures of from 6,000 to 7,500 K and are almost twice the size of the Sun with six times its luminosity. They make up 3.03% of the main sequence stars and include Procyon, Canopus and Zeta Leonis.

- G-Type Stars, include the Sun, a typical G2 class star. They make up about 7.5% of main sequence stars and have temperatures of about 5000 to 6000K. Examples also include Capella, and Beta Aquila (in the constellation of the Eagle);

- K-Type stars are orange in colour and slightly cooler than the Sun. They make up about 12% of the main sequence stars and also include giant K-type stars such as Aldebaran, Pollux and Arcturus and some orange dwarfs, like Alpha Centauri B. There is a suggestion that K spectrum stars may potentially increase the chances of life developing on orbiting planets that are within the habitable zone.

- M-type stars are most common of stars in the main sequence (About 76%) but as they have very low luminosities (about 0.04 that of the Sun), few are bright enough to be visible to the unaided eye. Although most class M stars are red dwarfs there are also many giants and some supergiant such as VY Canis Majoris, and bright Antares and Betelgeuse. This class also includes some of the hotter brown dwarfs which are somewhere in a state between a large gas giant planet and a small star.

## 6.8 The Life-Cycle of Stars

H-R diagrams have been most useful in interpreting how stars are formed and how they die. Stars usually form because of a gravitational instability within a nebula of gas and dust, often triggered by shock-waves from nearby supernovae, explosion of a massive star. When this happens and there is sufficient mass within the region, it begins to collapse under its own gravity. As the collapse and the density increases, the gravitational energy converts into heat and the temperature rises until the temperature is enough to begin nuclear fusion of hydrogen and the proto-star forms, often surrounded by a rotating protoplanetary disk of dust and gas.

Figure 6.15: The Horsehead Nebula is a dark nebula (i.e. dust obscuring starlight in the background) which is in the constellation of Orion just south of the star Alnitak (Photo: NASA)

Figure 6.16: The Great Nebula in Orion is a bright nebula of luminous gas and dust – the birthplace of protostars (Photo: NASA)

214

Stars usually spend the majority of their existence as main sequence stars; however, stars of different masses have markedly different properties at various stages of their development. The ultimate deaths of stars are different depending upon their mass which also determines their luminosity:

Stars with masses less than half that of the Sun evenly distribute their helium as it is being produced from the fusion of hydrogen throughout the whole star while in the main sequence. Because of this, as they soon reach the end of their fuel of hydrogen. They do not have extra reserves and so, in theory, become **white dwarfs** which simply cool off after exhausting their hydrogen. Much of the helium is converted to carbon which fails to re-ignite and electrons are compressed into the smallest possible space available – often with masses similar to the Sun but a volume similar to that of Earth. However, as their lifetimes are extremely long, this process has not been observed.

Stars with masses about half the size of our Sun up to about two times its mass and including our Sun, are able to use a secondary fuel of hydrogen in a shell outside of the central core of helium when the main supply has been depleted. The star then expands and cools to form a red giant star. For our Sun, this will probably occur in about 5 billion years from now, when it will expand to a roughly 250 times its present size. When their hydrogen is completely depleted, they then re-ignite the helium core and become a white dwarf and expel some mass as a planetary nebula.

Very large stars with masses up to ten times that of the Sun become supergiants once they have exhausted their entire core hydrogen. Further fusion of heavier elements occurs and when their core finally collapses they explode as a supernova. When this happens, depending upon the mass which remains, the star may then collapse and form a small, dense **neutron star**. Composed mainly of neutrons, these stars may be only kilometres across but have a mass several times that of the Sun. Some neutron stars rotate very rapidly over several hundred times a second and in doing so emit beams of electromagnetic radiation as **pulsars**. If the original mass at the time of the supernova is very large, then the resulting collapse after the explosion may result in an extremely compact and dense neutron star with such a large gravitational pull that even photons from light will not leave its vicinity or event horizon and it is termed a **black hole**.

The remains of nova and supernova explosions become the large masses of dust and gas or nebulae which then become the birth-place of new stars as gravitational concentrations within them once more bring material together to form new stars.

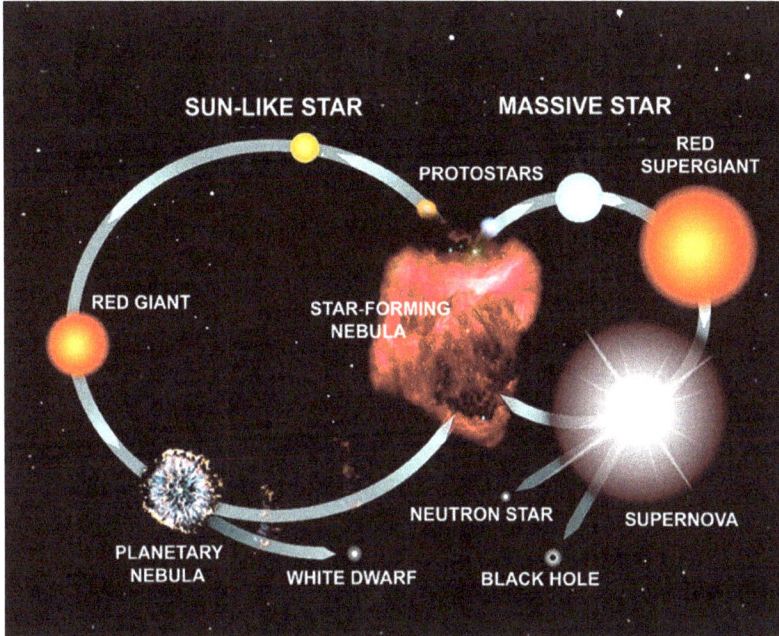

Figure 6.17: Diagram showing the life cycle of some stars. Massive stars have 8 to 10 times the mass of the Sun (Image: NASA)

## 6.9 Quasars

This abbreviation of **quasar** is derived from their full name of quasi-stellar radio sources, and they are very distant objects which are extremely luminous and are a source of intense amounts of electromagnetic radiation including radio waves. They have a very strong **red shift**, that is, they are moving away from us at great speed so that their light wavelengths appear to move towards the red end of the visible spectrum. This is a variation of the **Doppler effect**, first proposed by **Christian Doppler** (Austrian: 1803 – 1853) who observed that sources of sound waves which were moving away from an observer, produced wavelengths which became longer as the object move away and shorter if coming closer. For visible light, wavelengths moving away from us also become longer and move into the red end of the visible spectrum. Most stars have a red shift which can be measured and used to determine their velocities away from us.

Quasars appear to be similar to stars rather than to galaxies which can also look like stars with the unaided eye or small telescopes. However, their spectra contain broad lines unlike those from stars which have narrower bands and so the term quasi-stellar was applied to them. Their luminosity can be 100 times greater than that of the Milky Way galaxy and it is thought that most quasars were formed approximately 12 billion years ago by collisions of galaxies in which their central black holes merged to form either a very massive black hole or a double black hole system. The energy emitted by a quasar comes from mass falling onto the zone around the black hole called its accretion disc.

## 6.10 More About Galaxies

A galaxy named from the Greek *galaxias* for milky, is a very large system of stars, dust, and **dark matter** bound together by gravity. Galaxies range in size from just a few thousand stars to giants with over $10^{14}$ stars, each orbiting their own centre of mass. Galaxies are categorized according to their shape, which include elliptical, spiral, barred, oval and irregular types. The shapes of galaxies are influenced by their neighbours, and galaxies are also known to collide with each other and many galaxies also have black holes in their centres. Galaxies also are grouped together in clusters and superclusters with immense distances between them.

Figure 6.18: The Spiral Galaxy NGC-4414 which has a similar shape to our Milky Way Galaxy in the constellation of Coma Berenices about 55,000 light years in diameter and about 60 million light-years away from Earth (Photo: NASA).

Our galaxy, The Milky Way, seen as a bright, broad band across the night sky is a barred spiral galaxy with a diameter about 100,000–120,000 light years across. It estimated to contain about 200–400 billion stars but may have more. Our Solar System is located on one of the arms, called the Orion Arm, of the spiral about 27,000 light-years from the galactic centre. In the centre of this huge, flat spiral is an inner bulge containing an intense radio source, named Sagittarius A, which is likely to be a massive black hole. The rotational period of the Milky Way galaxy is about 240 million years at the position of the Sun. This constant rotational speed contradicts the classical laws of motion suggesting that much of the mass of our galaxy is not apparent and may consist of considerable dark matter. The Milky Way as a whole is moving at a velocity of approximately 600 km per second with respect to extragalactic frames of reference. The oldest stars in the Milky Way are nearly as old as the universe itself and thus must have formed shortly after the formation of the universe. The Milky Way also has several satellite galaxies, the Magellanic Clouds, and is part of the **Local Group** of galaxies, a component of the **Virgo Supercluster**, which is itself a component of the **Laniakea Supercluster** - Hawaiian: Laniakea, meaning immeasurable heaven.

Figure 6.19: Long-exposure image from NASA's Hubble Space Telescope of massive galaxy cluster Abell 2744 (Photo: NASA/ESA/STScI)

## 6.11 The Universe and Cosmology

The Universe includes planets, stars, galaxies and all matter and energy that has existed since it began at best calculation of $13.799 \pm 0.021 \times 10^9$ years ago. So far, what can be observed covers about 28 billion parsecs (91 billion light-years) in diameter but the size of the whole universe is not known and may be infinite. **Cosmology** named from the Greek *kosmos* for *world* and – *logia for* study, is the study of the origin, evolution, and eventual fate of the universe.

At the turn of the 20thy Century, most astronomers believed that the Milky Way was the limit of all that existed in the night sky. Using the new Hooker Telescope at Mt. Wilson, **Edwin Hubble** (American: 1889 – 1953), identified a type of star which pulsated, varying greatly in both temperature and diameter and giving a well-defined stable period and amplitude - these were the **cepheid variables**, named from Delta Cephei in the

Figure 6.20: Edwin Hubble

constellation of Cepheus, named for the legendary King of Aethiopia and father of Andromeda. Whilst this star had been discovered in the 18th century, it was Hubble

Figure 6.21: Image from the Hubble Space Telescope showing the Cepheid Variable star RS Puppis (Photo: NASA).

who first observed in the period 1922-1923, that these stars and the nebulae which contained them, were much too distant to be part of the Milky Way and were, in fact, entire galaxies outside our own. Hubble recognized that the constancy of theses bright light sources seen through a good telescope could be used to estimate distances to these galaxies, many of which were beyond that possible using the standard method of parallax.

In 1929, Hubble examined the relation between distance and red shift of galaxies using his Cepheid technique and found a rough proportionality between these distances and their Red Shifts, nowadays termed **Hubble's law**. This

Law states that distant galaxies were receding away from the Earth and that their velocities were proportional to their distances from the Earth. That is, those galaxies which were further away had greater velocities than those which were closer. The reasons for this were still unclear at the time and it was **Georges le Maître** (Belgium: 1894 – 1966), a Belgian Catholic priest and physicist, who found that Hubble's observations supported the theory of an expanding universe based on the equations for General Relativity formulated in 1915 by **Albert Einstein** (German: 1879 – 1955).

In 1964 American radio astronomers **Arno Penzias** (1933- ) **and Robert Wilson** (1936 - ) accidentally detected a faint background noise during their work with cosmic radio sources and suggested that this noise was due to microwave radiation which was possibly a relic of that produced at the very beginning of the universe, as it was found in all parts of the sky observed by their radio telescopes. This **cosmic microwave background (CMB**) radiation became one of the cornerstones of the new theory of the universe – the **big bang theory** – a popular theory suggesting that our universe sprang into existence as a singularity (a point source) around 13.7 billion years ago and still continues to expand outwards.

Many scientists could not come to terms with a universe that had a finite beginning. They therefore advanced theories in which the universe was expanding at the present time, but didn't have a beginning but was constantly being renewed. One such theory was the **steady state theory** proposed by **Hermann Bondi** (Austrian/British: 1919-2005), **Thomas Gold**

(Austrian/British: 1920 - 2004), and **Fred Hoyle** (British: 1915-2001) in 1948. In this theory, as old galaxies moved apart, new galaxies were formed from matter being created throughout space. The universe was therefore constant and that this could be tested by observation. Unfortunately, radio telescope surveys showed that there were more radio sources in the past, rather than a constant evolution and this was contrary to the basic assumption of the steady state theory. It was therefore abandoned.

The current big bang theory suggests that when the universe was young, it was denser, much hotter, and filled with a uniform, diffuse cloud of expanding hydrogen plasma as charged nuclei of hydrogen atoms. As this universe cooled, the protons from the hydrogen plasma combined with electrons to form neutral hydrogen atoms and thus reducing the plasma cloud making the universe more transparent. Cosmologists refer to this time as the **recombination epoch**.

After this time, came a period when photons (light particles) began to move freely through space rather than being scattered by electrons and protons in the plasma cloud. This is referred to as **photon decoupling** and these photons and their increase in wavelength over time due to expansion within the universe is the source of the relic radiation. Observations made by optical telescopes in the 1990s could not account for the amount of the expansion nor the background radiation detected by radio telescopes, so it was suggested that these discrepancies were due to theoretical **dark energy** and dark matter which were normally hidden from traditional optical telescopes but which comprised a considerable amount of the energy and matter of the universe.

After the initial expansion, the cooling universe allowed the formation firstly of simple atoms then clouds hydrogen and possibly other simple elements, which later condensed into stars, most stars into galaxies, most galaxies into clusters, superclusters and, finally, large-scale galactic filaments. It is estimated that the observable universe contains approximately $3\times10^{23}$ stars and more than 100 billion galaxies. Between these structures are empty spaces, which are typically up to 490 million light years in diameter. However, the observable material of the galaxies of the universe did not seem sufficient to have enough gravitational attraction to hold them together. In 1933, **Fritz Zwicky** (Swiss-American: 1898-1974) suggested that the matter needed to produce the observable gravitational effects of galaxies was not seen as luminous nor illuminated light but was dark matter.

There have been many hypotheses about the formation and possible fate of the universe and cosmologists remain unsure about what preceded the Big Bang. It is unknown territory that many do not wish to enter. Some philosophers refuse to speculate and others rely on theology as an explanation. Some suggest that there may be many, parallel universes as a **multiverse hypothesis** that exists in the same time frame as our own.

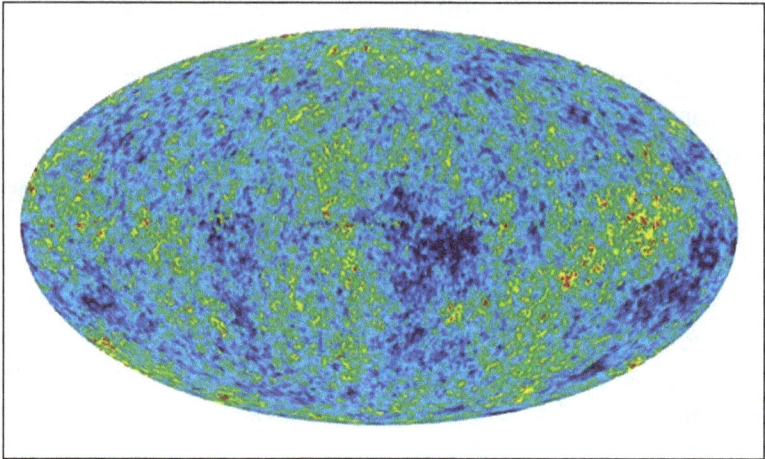

Figure 6.22: The detailed, all-sky picture of the infant universe created from nine years collected data showing 13.77 billion year old temperature fluctuations as colour differences corresponding to the seeds that grew into the galaxies (Photo: NASA / WMAP).

## 6.12 Seti - The Search for Extra-terrestrial Intelligence

It has been reasoned that if the universe contains approximately $3 \times 10^{23}$ stars, then surely an incredibly small percentage of these could have orbiting planets. Even a minute percentage would still give a probability of billions of potential planets which would be at such a distance from a suitable star to be in what is known as the **habitable zone** - that unique position which would possibly allow for the formation of life given that the planet itself has the right conditions of size and atmosphere.

There are extreme conditions on Earth in which living things have adapted to survive. It seems unlikely that there would be no planets in the universe that could sustain some form of life.

Figure 6.23: A small pool in the limestone caves at Jenolan, New South Wales, Australia deep underground without light and at a constant 15°C temperature which contains a small colony of springtails – primitive arthropods which are seen as wispy threads on the water's surface

Since 1988, over 2030 exoplanets or planets around stars other than our Sun have been discovered. The *Kepler Space Telescope* has detected a few thousand candidate planets and about 1 in 5 Sun-like stars have an Earth-sized planet in the habitable zone, with the nearest about 12 light-years distance from Earth. In February, 2017, NASA's *Spitzer Space Telescope* discovered the first known system of seven Earth-size planets around a single star with three of these planets firmly located in the habitable zone. Known as Trappist-1, the parent star is small and about 39 light years from Earth in the constellation of Aquarius. The planets are bunched together to take advantage of the star's dim light and heat with the nearest planet taking just 1.51 days to make one orbit, whilst the furthest planet takes 12.35 days. Whether or not any of these would have intelligent life is still to be determined, but it does give a good reason for searching for it.

**Nikola Tesla** (Serbian: 1856-1943) in 1899 thought that he had detected radio signals from Mars during his early experiments in wireless, but this proved to be an error in his work. Other pioneers in early radio and electricity, such as **Guglielmo Marconi** (Italian: 1874-1937) and **Lord Kelvin** (William Thomson: Irish/Scottish 1824-1907), also believed radio could be used to contact life on Mars. In 1924, with Mars at its closest position to Earth for a long while, a powerful radio receiver, tuned to a wavelength of between 8 and 9 kilometres, was sent aloft 3 kilometres in a US Naval dirigible balloon. The program was led by David Peck Todd (American: 1855 – 1939), a noted astronomer, but there was no positive result. By 1960, astronomer **Frank Drake** (American: 1930 - ) of Cornell University performed the first search for extra-terrestrial life as part of his *Project Ozma*, using a radio telescope of 26 metres diameter at Green Bank, West Virginia, to examine the stars Tau Ceti and Epsilon Eridani. He used a 400-kilohertz band single-channel receiver with a bandwidth of 100 hertz. He found nothing of interest.

In 1971, the National Aeronautics and Space Administration (NASA) funded a SETI study that proposed the construction of an Earth-based radio telescope array with 1,500 dishes known as *Project Cyclops*. But the cost was prohibitive and the project was cancelled. Only its report remained to act as an impetus for the further SETI work that followed.

On August 15, 1977, a SETI program from Ohio State University gave some hope when a project volunteer, **Jerry R. Ehman** (American: 1932 - 2006), noted strong signal received by their radio telescope on August, 15th 1977. His notation of "Wow" on the computer printout of the signal has become a by-word in SETI exploration and is so far, the best result for such research. Unfortunately it has never been detected again in later searches.

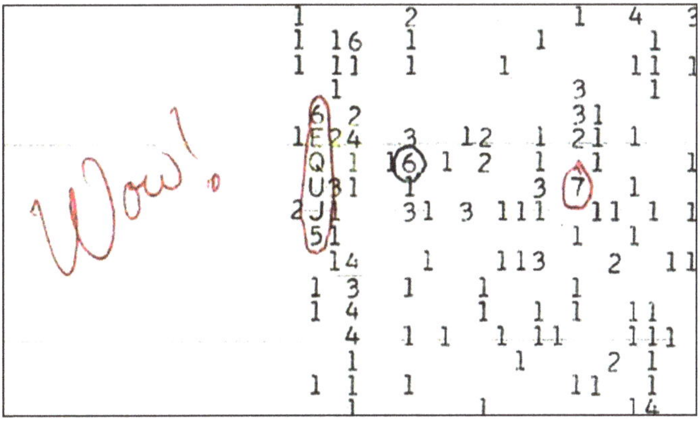

Figure 6.24: The Ehman WOW! notation (Photo: OSU and NAAPO, NASA)

In the 1980s, the Astronomer **Carl Sagan** (American: 1934-1996) and others founded the U.S. Planetary Society, partly as a vehicle for SETI studies, and several major projects were started in the United States. Many of these proposed the use of large radio telescope arrays and sophisticated computer-monitored receivers using a range of suitable wavelengths. Unfortunately the funding for these programs was considered to be a waste of money by the US Congress so funding was usually withdrawn. However, SETI advocates continued without government funding, and set up their own program called *Project Phoenix.*

From 1995 through to March 2004, *Phoenix* conducted observations at the 64 metre radio telescope at Parkes in New South Wales, Australia, the 43 metre radio telescope of the National Radio Astronomy Observatory at Green Bank, West Virginia, and at the 300 metre radio telescope at the Arecibo Observatory in Puerto Rico. The project studied about 800 stars over the frequency range from 1200 to 3000 Megahertz (MHz). Many international radio telescopes are currently being used for radio SETI searches, including the Low Frequency Array (LOFAR) in Europe, the Murchison Widefield Array (MWA) in Australia, and the Lovell Telescope in the United Kingdom. There has also been an on-going program called *SERENDIP* (Search for Extra-terrestrial Radio Emissions from Nearby Developed Intelligent Populations) since 1979 which allows for SETI-style searches during the usual on-going routine work at some radio telescopes at the University of California and at the Arecibo Observatory in Puerto Rico. There are also currently several public programs devoted to search and analysis of data, but to date there has been no positive evidence of life outside of Earth.

More recently, the world's largest single-dish radio telescope was launched in March, 2017 in China. The

Five-hundred-metre Aperture Spherical Telescope (FAST) has been built in a deep concave valley similar to that of the 305 metre wide Arecibo radio telescope in Puerto Rico. Built in the south-western province of Guizhou, this telescope uses a data system developed at the International Centre for Radio Astronomy (ICRAR) in Perth, Western Australia to manage the huge amounts of data it generates. It is hoped that the huge amounts of data which will be collected will allow astronomers to continue the hunt for new pulsars and continue the search for signals from extra-terrestrial intelligence.

Figure 6.25: Arecibo Observatory radio telescope in Puerto Rico (Photo: NASA)

# SUMMARY

1. The Earth is a small, rocky planet which is the only one so far known to support life. It orbits a medium-sized yellow main sequence star (the Sun) within the Milky Way galaxy.

2. The Earth and other planets were probably formed about 4.54 billion years ago as the Sun's solar nebula collapsed under its own gravitational pull to form a star with a spinning disk which accreted small clumps of material at different distances out from its centre to form the planets which, under the influence of the largest body Jupiter, then moved about until they formed into their current stable orbit.

3. The seasons of summer, autumn (fall), winter and spring are due to the Earth's axial tilt of about $23.5^0$ to the plane of orbit around the Sun. This causes the Sun's rays to strike the Earth's surface at different angles at different times of the Year.

4. The Moon appears from Earth to change its shape (or phase) during its 27 day cycle. It changes from a dark annulus of the new moon, through a crescent to the half shape of the first quarter, then a gibbous (partial) shape to a full or new moon then back again via an opposite crescent to the third quarter, another gibbous shape then back to a new moon again.

5. Tides are caused by the gravitational pull of the Moon and to a lesser extent, the Sun and there are usually two high tides and two low tides per day (semi-diurnal). When

the Sun, Moon, and Earth are aligned (syzygy), the tidal force of the Sun reinforces that of the Moon and gives a higher spring tide but when the Moon and the Sun are at 90° to each other, the Sun's tidal force partially reduces that of the Moon giving a lower neap tide.

6. Eclipses of the Moon  occur at might when the Earth passes between the Moon and the Sun such that the normally full moon will have the shadow of the Earth pass across it obscuring it completely or partially depending upon the relative positions of the three bodies. There can be partial and full eclipses.

7. Eclipses of the Sun occur during the day when the Moon passes between the Earth and the Sun and Moon completely (total eclipse) or partly (partial eclipse) covers the face of the Sun reducing daylight.

8. Early astronomers such as Aristarchus of Samos (310 – 230 BC), used detailed observations and crude angular-measuring devices to determine that the Moon was a sphere and at a known distance from the Earth. Later measurements were able to give a more accurate average for the distance to the Moon which is 362,600 km at perigee (closest position) and 405,400 km at apogee (furthest position) with an value of 384,400 kilometres.

9. The prevailing theory as to the Moon's formation, 4.5 billion years ago, is the Giant Impact Hypothesis which suggests that the Moon was formed by a glancing  impact of a large, Mars-sized  body (called Theia) impacting the proto-Earth not long after its formation. In addition, evidence from crater samples suggests that between

approximately 4.1 and 3.8 billion years ago numerous asteroid impacts during the Late Heavy Bombardment (LHB) episode caused significant changes to the surface of the Moon forming the craters, lava plains (mare), rilles and highlands.

10. Because the Moon is locked in a fixed orbit around the Earth, its rotational period is the same as its orbital period (27.3 days) so it always shows the same face to the Earth. The far side has more craters and fewer mare than the near side.

11. The Moon's interior is probably layered, with the surface layer being very thin and composed of anorthosites (plagioclase-rich rocks) and mafic rocks similar to those on Earth. Below this thin crust is a rocky mantle with a mafic and iron composition over a fluid outer core of iron and an inner core of partially molten metallic iron with a small amount of nickel.

12. Flying to the Moon has always been the dream of humankind and the subject of many science fiction works, but it was not until the invention of the liquid-fueled rocket and many successful and unsuccessful flights that humankind was able to first step onto the Moon's surface. This occurred on July 21, 1969, when Neil Armstrong of the Apollo 11 Mission took the first step.

13. The Solar System consists of the Sun, around which eight major planets (terrestrial planets such as Mercury, Venus, Earth and Mars; gas giants such as Jupiter and Saturn; and ice giants such as Uranus and Neptune), a

number of smaller (or dwarf) planets (including Pluto), asteroids, meteors, comets and much smaller debris left over when the Solar System was formed about 4.5 billion years ago. They all move at various speeds in separate elliptical orbits. Further out from the major planets is the Kuiper Belt, which contains the minor planets, many asteroids and frozen ices (frozen gases such as methane, ammonia and water). Beyond the Kuiper Belt and extending over the whole Solar System as a sphere of about 14,960 billion kilometres radius, is the Oort Cloud which is the outer limit of the Solar System.

14. The use of telescopes in astronomy began with Galileo's studies in 1609 and continued on through Newton's invention of the reflecting telescope in 1668 and on to more modern times with the application of radio technology and the invention of the radio telescope in the 1930's and the launching of space telescopes after 1990.

15. Stars, such as our Sun, are luminous spheres of plasma (very hot, electrically-charged gas) held together by strong gravitational pull which balance their expansion by heat. They give off radiating light, heat and other types of radiation including the solar wind. They are usually formed by collapsing masses of gas and dust within nebulae and are grouped together in large galaxies which themselves are part of larger clusters of galaxies.

16. Stars are of different colours depending upon their composition, size and brightness. Their brightness (luminosity) can be referenced as apparent (visual)

magnitude or absolute magnitude which has been scaled to a set distance (10 parsecs) for comparisons.

17. Hertzsprung- Russell (or H-R) diagrams are useful graphs of spectral (colour) class and luminosity and can be used to explain the evolution of different types of stars. Smaller stars become compact neutron stars of planetary size, but very large stars explode as nova and super nova producing large amounts of gas (mostly hydrogen) and dust for the formation of new stars.

18. Stars, galaxies, nebulae and all of the material and energy in space comprise the Universe which probably formed during the Big Bang about $13.8×10^9$ years ago when matter was created.

19. S.E.T.I (Search for Extra Terrestrial Intelligence) programs have been operating since the development of radio astronomy and continue today. This has involved using sophisticated computer scanning of a variety of frequencies of many star systems which have many exoplanets but to date there have been no indication of life other than on Earth.

# Practical Tips

1. Astronomy is one of the few sciences in which the skilled amateur can make an impact, especially in the study of comets and smaller planets. Even with a modest telescope, one can view many of the wonders of the night sky such as planets, their moons, nebulae, galaxies and multiple star systems.

2. The best telescopes to use are reflecting telescopes with parabolic mirrors as these can be made cheaper and larger than refracting telescopes. Because of the size of their aperture, reflecting telescopes have better light-gathering power (resolution) giving a clearer image. At least an 8 inch Newtonian or Cassegrain telescope is best for the new amateur. Telescopes can be mounted so that they can follow stars using computer-aided electric drives, but this requires careful setting up.

3. Observation is best done away from light pollution, preferably well away from cities to also reduce atmospheric pollution and not during periods of a full Moon. A good position well above trees and buildings is the best, and a view towards the ecliptic (e.g. towards the equator) will ensure that constellations and planets can be easily seen rising in the east and travelling across the sky to the west.

3. There are many useful astronomy apps. which can be used to locate various objects in the night sky. Winter is best for observation because of the stability of the colder atmosphere but one may have to wait until the early hours to locate the desire object.

4. Constellations are a good target for searching, as they are given as the main reference for night sky objects. Some constellations can be easily recognized, such as Orion, Scorpio, Ursa Major and Canis Major, but others require the use of charts or astronomy Apps. Some constellations can only be seen in one hemisphere.

5. The Moon is a great source of study and there are good charts and Apps. to identify surface features. Whilst a full Moon washes out much of the other objects in the sky, it is best for observation of its surface features but a Moon filter or piece of Polaroid is needed to reduce the glare.

# Multichoice Questions

1. Many planets have a tilted axis of rotation suggesting that:

    A. these tilts were caused by eddy currents at formation
    B. they suffered impacts early in their formation
    C. they were caused by the gravitation attraction of their moons
    D. they had their poles attracted to the Sun

2. The very large clouds of dust and luminous gases in space are called:

    A. Nebulae
    B. Galaxies
    C. Constellations
    D. Black Holes

3. The brightness of a star if it were calculated as though the star was brought to a distance of 10 parsecs from the observer is called the star's:

    A. Luminosity
    B. Relative Magnitude
    C. Absolute Magnitude
    D. Distance Modulus

4.    Which of the following diagrams best shows the positions of the Sun, Moon and Earth for a neap tide:

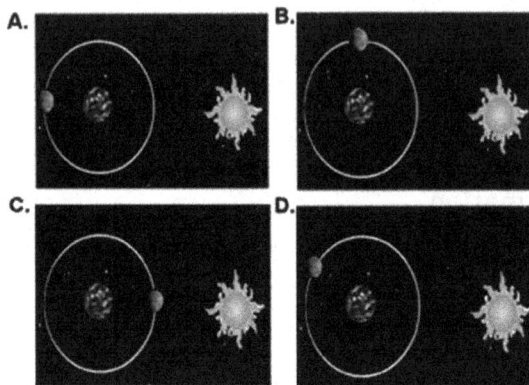

A.   A
B.   B
C.   C
D.   D


5.    The following information is given for a small reflecting telescope:

diameter of primary mirror=10 cm.
focal length of primary mirror=100 cm.
length of telescope = 90 cm.
light path distance =110 cm.
diameter of eyepiece =5 cm.
focal length of eyepiece = 2 cm.
height of image at 10 cm from eyepiece = 5 cm.

The magnification of this telescope is most likely:

A. 20
B. 50
C. 100
D. 150

6. An astronomer searching for new planets within the Solar System using a telescope and camera on Earth would most likely use as a control:

A. the background of fixed stars
B. the rotation of the stars seen from Earth
C. the positions of the other planets
D. the size of the developed photos

7. Tycho Brahe's long term observations assisted in the development of:

A. the inverse square law
B. Kepler's laws of planetary motion
C. the gravitational constant
D. the refracting telescope

8. A refracting telescope has a length of about 200 cm with a diameter of the objective lens of 150 mm. What is the resolution of this telescope?

A. 0.03 secs of arc
B. 0.12 secs of arc
C. 0.58 secs of arc
D. 0.77 secs of arc

9. The following diagram is a Hertzsprung-Russell diagram showing the luminosity and magnitude of stars plotted against several other parameters:

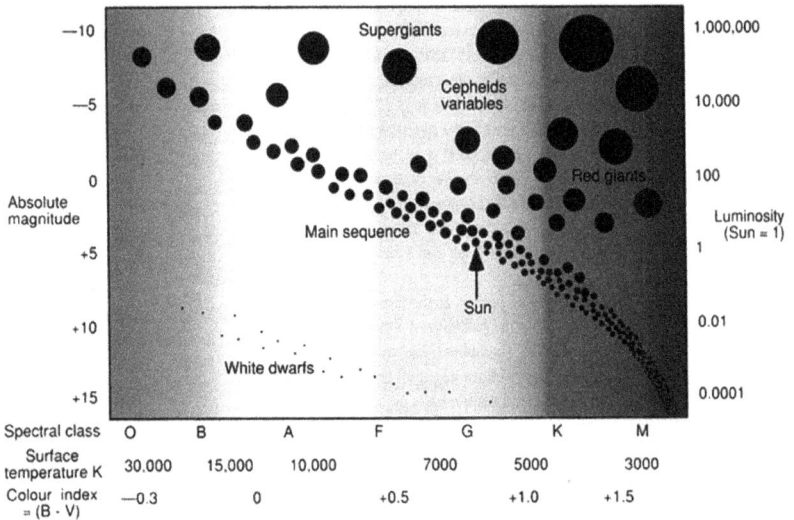

From this diagram it could be inferred that a star of the Main Sequence of Spectral Class A would :

A. be a white dwarf
B. have a surface temperature of 1500 degrees K
C. have an absolute magnitude with a negative value
D. have a luminosity similar to our Sun

10. This question refers to the following photograph of the surface of the Moon:

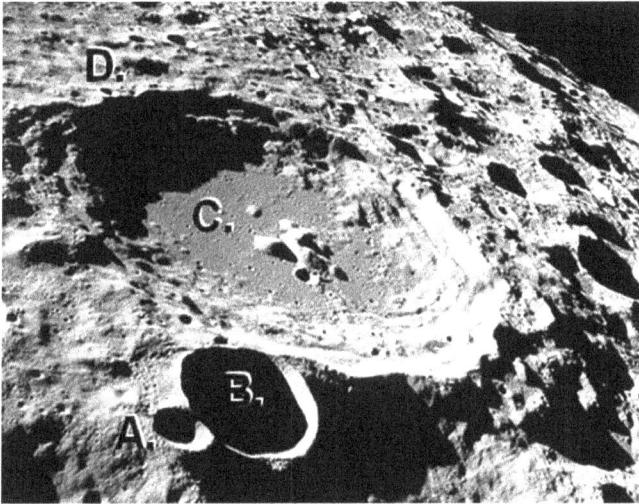

The youngest_crater shown here is:

A.  A
B.  B
C.  C
D.  D

## Review and Discussion Questions

1. Explain the differences between each of the following:

   (a) nebulae and galaxy
   (b) constellation and asterism
   (c) declination and right ascension
   (d) azimuth and zenith
   (e) resolution  and magnification
   (f) parsec and light year
   (g) asteroid and comet
   (h) terrae and maria

2. What was Bode's law? How accurate was it and was it of any use in astronomy?

3.  Explain how absolute magnitude is defined. Why was there a need to do this?

4.  State (giving the meanings of symbols and units):

   (a)  Kepler's three laws of planetary motion
   (b) Newton's law of universal gravitation
   (c) the formula for Dawes' limitation in resolution
   (d) the inverse square law

5.  Compare  and  contrast  the  advantages  and disadvantages  between  reflecting  and  refracting telescopes.

6. Discuss the requirements that would be needed for establishing a permanent colony on:

   a. the Moon
   b. Mars

Give a brief outline of how these requirements could be met.

7. Explain how spectral analysis of starlight has been used to determine the speed and relative motion of stars.

8. The following diagrams represent separate spectrograms showing bright lines of different colours of the visible spectrum on a black background as seen through a spectroscope attached to the end of a reflecting telescope:

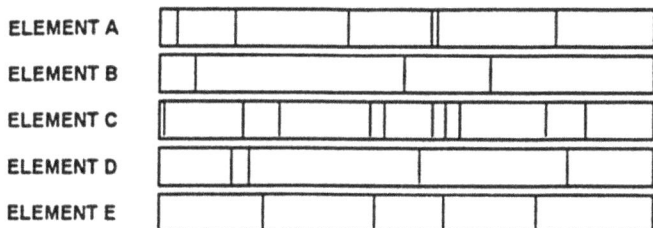

Spectrum seen when STAR X is observed.

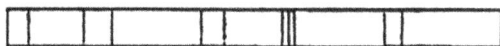

a. From the data above, list the element or elements which predominate on Star X

b. Explain how this type of spectrum differs from that which is usually observed from the Sun by referring to the main types of spectra (internet may help).

c. The astronomer thinks that this telescope/spectroscope system has an error and that the colours observed are slightly wrong.

Design an experimental test which could be used to test this system to see if there is any error in the equipment.

9. Given that Kepler's Third Law states that period squared is proportional to distance cubed, use a constant (k) for this expression from Earth data (distance = 1 AU and period = 1 year) to calculate the distance from the Sun of another planet having a period of 84 Earth years.

10. Use the Internet to review current SETI programs.

**Answers to the Multichoice Questions**

Q1. B  Q2. A  Q3. C  Q4. B  Q5. B
Q6. A  Q7. B  Q8. D  Q9. C  Q10. B

# Reading List

Bussey, B., Spudis, P.D. (2004). *The Clementine Atlas of the Moon.* Cambridge University Press. ISBN 0-521-81528-2.

Canup, R.M. & Asphaug, E.(2001). *Origin of the Moon in a giant impact near the end of the Earth's formation.* Nature 412: pp.708-712. Bibcode: 2001 Nature412708C.doi:10.1038/35089010. PMID 11507633.

Cattermole, Peter & Moore, Patrick (1985). *The Story of the Earth.* Cambridge: Cambridge University Press. ISBN 978-0-521-26292-7.

Dalrymple, G. Brent (2001). *The age of the Earth in the twentieth century: a problem (mostly) solved.* Special Publications, Geological Society of London 190 (1): 205-221. Bibcode: 2001GSLSP. 190205D

D'Angelo, G.; Marzari, F. (2012). *Outward Migration of Jupiter and Saturn in Evolved Gaseous Disks.* The Astrophysical Journal 757 (1): 50. (23 pp.). arXiv: 1207.2737. Bibcode: 2012ApJ75750D.doi:10.1088/0004-637X/757/1/50.

Eddy, J. (1979). *A New Sun: The Solar Results From Skylab.* NASA. p. 37. NASA SP-402.

Izidoro, A., Haghighipour, N., Winter, O. C. & Tsuchida, M. (2014). *Terrestrial Planet Formation in a Protoplanetary Disk with a Local Mass Depletion: A Successful Scenario for the Formation of Mars.* The

*Astrophysical Journal* 782 (1): 31, (20 pp.). arXiv:1312.3959. Bibcode:2014ApJ78231l.doi:10.1088/0004-637X/782/1/31.

Lang, Kenneth R. (2011), *The Cambridge Guide to the Solar System*, 2nd ed., Cambridge University Press.

*Lubow, S. H. & Ida, S. (2011). Planet Migration. In Seager, S. Exoplanets. University of Arizona Press, Tucson, AZ. pp. 347-371.* arXiv:1004.4137. Bibcode:2011exop.book347L

*Lucey, P.; Korotev, Randy L.; et al. (2006). Understanding the lunar surface and space-Moon interactions. Reviews in Mineralogy and Geochemistry 60 (1): pp.83-219. Doi: 10.2138 /rmg 2006.60.2.*

Malville, J. McKim, Thomson, Hugh & Ziegler, Gary.(2004). *Machu Picchu's Observatory: the Re-Discovery of Llactapata and its Sun-Temple.* (Longer English version of the article that was first published in the Revista Andina ,2004, #39).

*Moore, Patrick & Rees, Robin. (2014). Patrick Moore's Data Book of Astronomy. Cambridge, U.K. Cambridge University Press.* ISBN 1139495224.

Munns, David P.D. (2013). *A Single Sky: How an International Community Forged the Science of Radio Astronomy.* Cambridge, MA: MIT Press

NASA: https://www.nasa.gov/

Singh, S. (2005). Big Bang: The Origin of the Universe. Harper Perennial. pp. 560.

Sridharan, R., Ahmed, S.M., Dasa, Tirtha Pratim, Sreelathaa, P.,Pradeepkumara, P., Naika, Neha & Supriya, Gogulapati. (2010). Direct' evidence for water (H2O) in the sunlit lunar ambience. From CHACE on MIP of Chandrayaan I. Planetary and Space Science, 58 (6). pp. 947-950. ISSN 0032-0633

Stroud, R. (2009). The Book of the Moon. Walken and Company. pp. 24-27. ISBN 978-0-8027-1734-4.

United States Geological Survey (USGS). (2008). Dwarf Planets and their Systems". Working Group for Planetary System Nomenclature (WGPSN). U.S. Geological Survey. 2008-11-07.

Williams, J. (2010). The astrophysical environment of the solar birthplace.Contemporary Physics 51 (5): p.381-396. arXiv:1008.2973. Bibcode:2010ConPh..51..381W. doi:10.1080/00107511003764725

Sullivan, Woodruff T. III. (2009). Cosmic Noise: A History of Early Radio Astronomy. Cambridge University Press,

Zeilik, Michael A. & Gregory, Stephen A.(1998). Introductory Astronomy & Astrophysics (4th ed.). Saunders College Publishing. ISBN 0-03-006228-4.

## Some Useful Free Astronomy Apps:

*SkEye* – *an advanced app that can be attached to telescope mounts which allow   control  (PUSHTO).  4.3. rating*

*Night Sky Tools* – *an excellent app. OFFLINE with red vision for night use, wide range  of  data  on  eclipses, Moon, Planets and stars*

*Skymap* – *good for charts of the night sky and planets. Rated 4.5;*

*Star Walk* – *rated at 4.3 and has some in-house purchases. Good 3D Models and also   gives   artificial satellites, dwarf planet missions.*

*Star Chart* – *4.2 rating is a good basic star map.*

*Planets' Position 4.2  good for planets and the Moon;*

*Moon Book 4.1 is useful for eclipse data;*

*Eclipse Calculator 4.2 gives solar and lunar eclipses. Also available in Spanish.*

*Lunar Eclipse 3.4;*

*Solar Walk 4.0 for Solar System data and 3D images. Some in-house purchases.*

*NASA App 4.3 for the latest from NASA. NASA TV.*

# Key Terms Index
## (Page numbers in brackets)

**absolute magnitude M (208)** is a measure of the brightness of a star which includes a distance factor by converting all values as though the star was at a standard distance of 10 parsecs (i.e. 32.6 light years).

**acceleration (75)** is the increase in velocity per time (e.g. km/sec/sec i.e. $km/sec^2$).

**accretion (7)** is the gathering up of smaller material as a larger object moves through it with the subsequent attachment of the matter to the larger object which then grows in size.

**achromatic lenses (93)** are compound lenses which have an additional correcting lens joined with others to reduce dispersion.

**achrondrites (180) are stoney meteorites** which do not contain.

**altitude (187)** is the angular height of the object from the horizon (from $00^0$ at the horizontal to $90^0$ at vertical) used in communicating height of a celestial body in the sky.

**annulus (18)** is a ring such as when the new moon appears like a faint ring with a darker centre.

**anorthosites (33)** are a family of igneous rock which have over 90% plagioclase composition and are found on the surface layer of the moon as well as on Earth.

**aperture (88)** of telescopes is the opening which gathers the light to be magnified. For refractors this is the area of the objective lens and for reflectors this is the area of the primary mirror. The bigger the aperture, the clearer is the image.

**aphelion (2)** is the furthest distance that a planet is from the Sun.

**apogee (13)** is the furthest distance that the moon moves away from the Earth in its orbit.

**apparent magnitude (206)** also called visual or relative magnitude was an early scheme of rating stars from bright (first magnitude) to dull (sixth magnitude) depending on how bright they appeared. However, astronomers soon realized that there were stars much bright than 1 and fainter than 6, and so a more sophisticated system was needed.

**asterism (201)** is part of a constellation which seems to have its own separate pattern which is useful in identification e.g. the big dipper is part of Ursa Major and is used to find the star Polaris and North.

**asteroids (177)** are small to large, irregular bodies which orbit the Sun in a variety of orbits.

**asteroid belt (146)** contains the orbits of very many smaller pieces of rock and/or ice orbiting the Sun. There are near Earth asteroids and Trojan and Greek asteroids which follow around Jupiter in its orbit at Lagrange points.

**astrolabe (20)** is a simple device for measuring angles which later developed into the sextant.

**astronomical unit AU (108)** is a unit of distance taken as that distance between the Earth and the Sun as 149,597,870,700 metres.

**astronomy (1)** from the Greek *astron* for the stars and - *omy* for study of is one of the oldest of the sciences and is the study of the night sky.

**azimuth (187)** is the geographical bearing $000^0$ to $360^0$ (with North at 0000, east at $090^0$ and South at $180^0$ and so on) used in communicating direction of a celestial body in the sky.

**barycentre (14)** the central point or centre of mass of the orbit of one object around another.

**big bang theory (227)** is the current theory to explain the formation of the universe about $13.8\times10^9$ years ago, in which the universe was created by a sudden burst of energy forming plasma which cooled and condensed into matter and eventually the celestial objects observed today.

**binary star system (198)** consists of two stars which are orbiting around their centres of mass. There are also triple systems e.g. Alpha Centauri.

**black hole (220)** is a region around a very dense neutron star in which nothing, including light photons, can escape the strong gravitational pull. The boundary of the region from which no escape is possible is called the event horizon.

**bolloids (179)** are larger meteoroids which burn up in the Earth's atmosphere. They may be large enough to be seen during the day.

**capture theory (24)** suggests that the Earth's gravity may have captured a passing body which became its moon during the early time of the solar system's formation.

**Cassegrain reflector (89)** is a reflecting telescope which uses a parabolic primary mirror with a convex secondary mirror to re-direct and focus the light through the rear of the telescope, through the primary mirror to the eyepiece.

**catadioptric telescope (89)** is another name for Cassegrain reflectors because they use a combination of corrected lenses and mirrors to obtain a very clear image.

**celestial sphere (186)** is an imaginary sphere around the Earth containing coordinate lines used for plotting the locations of heavenly objects.

**Centaurs (178)** are minor planets existing between the outer planets and have the characteristics of both asteroids and comets.

**centrifugal force (79)** is Latin for centre fleeing and describes the tendency of an object following a curved path to fly outwards away from the centre of the orbit.

**cephid variables (226)** are types of stars which pulsate, varying greatly in both temperature and diameter and giving a well-defined stable period and amplitude.

**chromatic aberration (92)** is the lens defect giving coloured edges to the image due to the imperfect grinding of the lens. Often found in cheaper, small aperture refractors.

**chromosphere (107)** is the atmosphere of the Sun just above its surface.

**chrondrites (180) are stoney meteorites** which contain chondrules or round grains or droplets of minerals.

**co-accretion theory (23)** suggests that moons can also form at the same time as their parent planet.

**coma (182)** is the diffuse haze which surrounds the central head of a comet and which may stream out as a tail when the comets comes close to the Sun and is pushed out by the solar wind.

**comets (182)** are large masses of ices which have very elliptical orbits and come in from deeper reaches of the solar system and are seen only when close to the Sun when their surfaces are partly vaporized giving a tail which is blown away by the solar wind.

**conjunction (118)** the alignment of two bodies as seen from Earth in relationship to the Sun.

**constellations (117)** are patterns of stars seen in the night sky.

**convective zone (106)** of the Sun is where heat is transferred to the surface via convection currents.

**core accretion model (6)** is the current theory of the formation of the solar system and was based upon the earlier nebular hypothesis. This suggests that a huge cloud of dust and gas, mostly hydrogen and helium, known as a solar nebula, collapsed under the influence of its own gravitational pull to form the Sun and its surrounding objects.

**corona (107)** is the very hot glow around the Sun which extends a vast distance out into space.

**cosmic microwave background CMB (227)** microwave radiation which was possibly a relic of that produced at the very beginning of the universe.

**cosmology (225)** is the study of the universe and of its formation.

**craters (27)** are large, circular impact structures.

**crater rays (29)** are shown as bright, streaks radiating out from some of the more prominent craters.

**chromatic aberration (92)** in lenses is caused by poor shape which breaks up light into a spectrum giving coloured edges to images.

**dark energy (229)** suggested unobserved energy which permeates the universe and assists in the acceleration of its expansion.

**dark matter (223)** is the matter which is unobserved and thought to account for much of the gravitational effects that is observed in the universe.

**Dawes limit (91)** is the resolving power limitation of any telescope and is given by r =116/d where r is the resolution (in seconds of arc) and d is the aperture size of the telescope (in millimetres).

**declination (186) Dec** is measured in degrees and minutes and seconds of arc which indicate how far north or south the sky object is from the celestial Equator, the projection of the Earth's Equator onto the celestial sphere.

**density (4)** is the mass of an object divided by its volume. for comparison, the density of water is taken as a standard of 1 gram per cubic centimetre.

**diffraction (91)** is the bending of light around an edge giving distortion in telescopes with small apertures.

**dioptre (90)** is a term for the power of a lens and is equal to 1/focal length measured in metres.

**dispersion (20)** is the breakup of white light (i.e. Sunlight) into its component colours by the atmosphere, raindrops or badly cut lenses.

**distance modulus μ (208) is** the difference between apparent and absolute magnitudes as a way of calculating distance to a star.

**diurnal tides (13)** have one high and one low tide per day.

**Doppler effect (222)** as sources of waves move away from an observer their wavelengths which became longer as the object move away (and shorter if coming closer). For visible light, wavelengths moving away from us also become longer and move into the red end of the visible spectrum (the red shift).

**dwarf planets (172)** are those small planets which orbit the Sun, are massive enough and are nearly rounded in shape, but still shares its orbit with a considerable amount of debris e.g. Ceres and Pluto.

**eclipse (19)** occur when one body passes between the viewpoint and another body such as the eclipse of the Moon when the Earth passes between the Sun and the Moon.

**ecliptic (100)** is the pathway which the Sun, planets and stars seem to make around the Earth because of the Earth's motion in the flat plane of the Sun and its solar system.

**elongation (119)** is the angle between the Sun and a planet often used when discussing the altitude or height above the horizon, of planets closer in to the Sun (e.g. Venus and Mercury).

**epicycles (68,132)** are circles within a circular orbit. Mars appears to make loops inn the night sky and this was once the (incorrect) model for its orbit.

**Equatorial coordinate system (186)** is a system used to align telescope mountings and is the projection of geographical latitude and longitude onto the celestial sphere, an imaginary sphere encircling the Earth.

**equinoxes (9)** occur twice the year when the Sun is directly overhead at the Equator.

**exoplanets (104)** are planets which are found outside of our solar system.

**focal length (85)** is the distance between a lens and its focal point.

**force (75)** a push or a pulling action which changes the position of a mass.

**fusion (107)** is the nuclear reaction in which the centres (nuclei) of smaller atoms such as hydrogen are forced together to become larger nuclei such as helium with the expulsion of vast amounts of light, heat, neutrons and other radioactive particles.

**galaxy (100)** is a very large group of stars, dust and gas. Our galaxy, the Milky Way, is a typical spiril galaxy but others can be barred or spherical.

**gas giant (148)** is mainly composed of hydrogen with a little helium, but it may also have a rocky core of heavier elements but lacks a well-defined solid surface e.g. Jupiter and Saturn.

**geocentric theory (66)** was an old theory suggesting that all the heavingly bodies orbited the Earth.

**giant impact hypothesis (24)** is the prevailing theory suggesting that the moon was formed by a glancing blow of a large body (Theia), impacting the photo-Earth not long after its formation. At impact, much of this body and perhaps part of outer layer of the proto-Earth were thrown into space, accreted together in orbit around the larger Earth forming the Moon.

**gibbous (18)** is an oblique or slightly off spherical shape.

**gravitation (76)** is the force of attraction between any two masses.

**Great Dark Spot (168)** was a huge rotating storm in the southern atmosphere of Neptune about the size of the entire Earth with winds measured at speeds over 2000 km/hr. it was first discovered when the voyager 2 spacecraft flew by Neptune in 1989 but when the Hubble space telescope looked at Neptune in 1994, the Great Dark Spot was gone.

**Great Red Spot (149)** is a giant, rotating storm in Jupiter's atmosphere more than twice the size of Earth. Winds inside this storm reach speeds of over 400 km/hr.

**Greeks (83)** are Trojan asteroids which are at the l4 Lagrangian point in front of Jupiter.

**habitable zone (231)** is that exact distance of a planet from a star of suitable size and energy which would allow life to exist on the planet.

**Harvard spectral classification system (214)** originally used capital letters running alphabetically to denote the colour of stars, but was later reordered to reflect their surface temperatures. in order of decreasing temperature, these types were o, b, a, f, g, k. and m.

**heliocentric theory (66)** is the modern theory that the Earth and all of the planets orbit the Sun.

**Hertzsprung- Russell (or h-r) diagrams (213)** are graphs of a factor of a star's colour (or temperature or spectral class) against its brightness (luminosity or absolute magnitude)

**horizontal coordinate system (187)** is a more casual of locating night objects using azimuth (the geographical bearing 000° to 360°) and altitude between observers who are at about the same location and observing at about the same time.

**Hubble's law (227)** states that distant galaxies are receding away from the Earth and that their velocities are proportional to their distances from the Earth.

**ice giant (163)** planet which is mostly composed of greater percentages of ices, mainly of water, methane and ammonia surrounding a smaller core e.g. Uranus and Neptune.

**ionisphere (46)** is an upper layer of the Earth's atmosphere which is rich in ions or charged particles which often cause interference in Earth-based radio.

**inferior planet (118)** is one between the Earth and the Sun i.e. Venus and Mercury.

**interferometers (97)** are devices which can take several wave signals and combine them into one pattern allowing them to be analyzed as one signal.

**inverse square law (71)** for radiation states that the intensity of this radiation varies inversely (i.e. if one factor goes up, the other goes down) with the square of its distance from the radiation source.

**isotope (108)** is a different form of a particular element having different numbers of neutrons within its nucleus e.g. deuterium or hydrogen-2 has an extra neutron in its nucleus whereas normal hydrogen has no neutron.

**kelvin (107) is a scientific temperature scale** where 1 K = -273 Celsius degrees - the temperature of absolute zero where ideal gases would contract to zero volume. The unit is named after William Thomson, 1st Baron Kelvin (Scots-Irish: 1824-1907).

**Kuiper belt (100)** is a zone of minor planets, asteroids and smaller bodies of ices which orbit the Sun well beyond that of the planet Neptune.

**Lagrange points (95)** are distinct positions between bodies such as the Sun and Earth, or within a planet's orbit where another object could occupy in a gravitationally stable position.

**Laniakea Supercluster (224)** from the Hawaiian name: Laniakea, meaning immeasurable heaven, is the extremely large group of galaxies containing those known.

**late heavy bombardment episode LHB (25)** occurred between approximately 4.1 and 3.8 billion years ago and explains the craters on the moon, suggesting that there were numerous asteroid impacts during this time.

**laws of motion (74)** devised by Sir Isaac Newton explains how objects act when forces are applied to them.

**laws of planetary motion (69)** devised by Johannes Kepler states three principles explaining how the planets orbit the Sun.

**laws of universal gravitation (77)** devised by Sir Isaac Newton states that the force of gravity between any two masses is proportional to the masses but inversely proportional to the square of the distance between them.

**light year (116)** a light year (LY) is the distance that light travels in one year at a velocity of $3.00×10^8$ m/s or about of $9.4607×10^{12}$ km.

**Local Group (224)** is the name given to the cluster of galaxies nearer to and including our Milky Way galaxy.

**luminosity (207)** is the total amount of energy emitted by an astronomical object per unit time and is measured in joules per second or watts but sometimes can be given as a comparison value to the luminosity of the Sun ($L_\odot$) which is $3.846×10^{26}$ watts.

**magnetosphere (151)** is the cavity created in the solar wind by the planet's magnetic field.

**magnification (90)** of any telescope is be found by dividing the focal length of the primary lens or mirror by that of each eyepiece.

**main-sequence stars (116)** refer to those stars on a Hertzsprung - Russell diagram which comprise about 90% of observed stars and are characterized by their generation of energy by the nuclear fusion of hydrogen to helium.

**maria (27)** are the flat, lava field on the Moon's surface.

**meteoroids (179)** is the general name applied to all fragments of rock or metal between dust size and that of small asteroids which come into the Earth's atmosphere.

**meteorites (179)** are meteors which strike the Earth's surface.

**meteors (179)** this is a general name for the smaller meteoroids which burn up in the Earth's atmosphere.

**momentum (76)** is defined as that possessed by a mass with a velocity.

**Morgan-Keenan system (214)** is a refinement of the Harvard system of classifying stars which uses alphabetical letters. in this system, each spectrum letter class has been expanded by a number from 0 to 9 and luminosity is expressed by the roman numerals i, ii, iii, iv and v, to indicate the width of certain absorption lines in the star's spectrum.

**multiverse hypothesis (230) suggests that there may be several universes co-existing.**

**neap tides (12) occur** when the Sun and moon at right angles giving high tides lower than is usual.

**nebulae (184)** are large masses of gas (often luminous) such as hydrogen and helium as well as dust in space. They are often the birthplace of stars. They can be bright nebulae with much luminous stars and gas or dark nebulae where much of the luminosity is obscured by dust.

**nebular hypothesis (6)** is an earlier concept suggesting that the solar system formed from the collapse of a giant cloud of interstellar matter within our galaxy, possibly triggered off by the explosion or supernova of a giant star nearby.

**neutron star (220)** is formed at the end of a very large star's life cycle when all of its fuel has been used and it collapses to a small body mainly composed of neutrons. it may be only kilometres across but have a mass several times that of the Sun.

**Newton's theory of colour (83)** suggested that colour is a property intrinsic to light and that colour is the result of objects interacting with already-coloured light rather than non-luminous objects generating the colour themselves.

**nice hypothesis (147)** is named after the French city and pronounced *neece* and suggests that the inner zone of rock was formed in place and when heated by the Sun, much of the ice and other volatile material was driven off.

**nova (191)** from the Latin for new and specifically referring to a new star, is a sudden brightening of a star as it explodes.

**nuclear fusion (6)** is the joining together of sub-atomic nuclei (centres having protons and neutrons) at high temperatures to produce new chemical elements and a tremendous amount of energy and radioactive particles. The most common occurs in stars where hydrogen nuclei are fused to for helium nuclei and larger atoms.

**Oort cloud (100)** is an immense sphere of dust, rock and ice which surrounds the solar system at about 15,000 billion kilometres, forming its outer limit.

**opposition (120)** is the alignment of the Earth and a planet away from the Sun.

**parallax (206)** is a method of calculating distance to stars and other objects using angles seen from different observation points.

**parsec (206)** is the distance at which one astronomical unit subtends an angle of one arcsecond and is equal to about 3.26 light-years.

**penumbra (19)** is the lighter, partial shadow on the edges of the darker part (umbra) during an eclipse of the Moon.

**perigee (13)** is the closest distance that the Moon comes to the Earth in its orbit.

**perihelion (2)** is the closest distance that a planet approaches the Sun.

**period (3,71)** is the time taken for an action to occur e.g. orbital period is the time taken for an object to go around its primary once.

**phase (18)** is the shape shown by an orbiting body in respect to the Sun and the observer e.g. the phases of the Moon and Venus.

**photon decoupling (229)** a suggested time when photons (light particles) began to move freely through space rather than being scattered by electrons and protons in the plasma cloud.

**photosphere (106)** is the visible surface of the Sun.

**planet (2)** illuminated body which orbits a star and contains most of the mass of material in that orbit.

**planetary migration hypothesis (105)** suggests that many of the larger planerts were formed close in to the Sun but due to their interaction and collision, they were propelled outwards to their current orbits.

**planetesimals (104)** are the small to large pieces of debris which orbited new stars and which may accrete into planets or gather as asteroids in belts around the star.

**plasma (106)** is the hot, electrically-charged material made up of the nuclei of atoms in stars. It is also seen as the flow of charged particles (electrons) in electrical discharges such as lightning.

**precession (9)** is the slight wobble of the Earth's axis.

**protoplanetary disc (102) is a flattened, rotating disc around the mass of a new star, which contains the material which will eventually accrete into planets.**

**pulsars (220)** are stars which are composed mainly of neutrons, are only kilometres across, have a great mass several times that of the Sun and rotate emitting beams of electromagnetic radiation.

**quadrature (119)** occurs when a celestial object makes a right angle with respect to the direction of the Sun.

**quasars (221)** or quasi-stellar radio sources, are very distant objects which are extremely luminous and a strong source of electromagnetic radiation.

**radiative zone (106)** is the interior zone of the Sun of uniform rotation of the internal plasma where the energy in the core is transmitted outwards.

**radio telescope (96)** is a sophisticated radio aerial, receiver and signal converter which detects the faint radio waves emitted by stars and other objects in space.

**Rayleigh scattering (20)** is the dispersion of light due to very fine particles such as those in the atmosphere.

**recombination epoch (229) is** a suggested time when the universe cooled, and protons from the hydrogen plasma which previously existed combined with electrons to form neutral hydrogen atoms, reducing the plasma cloud and making the universe more transparent.

**red giant (117)** are those stars at the end of their life cycle which have suddenly expanded as they generate new energy from the fusion of heavier elements.

**red shift (221)** is the apparent change in the colour of a star towards the red end of the visible spectrum as it moves away from the observer.

**reflecting telescopes (87)** use a curved (usually parabolic) primary mirror of reflecting glass, ceramic or metal to obtain the image and a glass eyepiece to magnify it.

**refracting telescopes (85)** are those which give a magnified image using only lenses – an objective lens at the front and an eyepiece at the other end.

**resolution (89)** refers to the clarity of the image seen with a telescope and is the ability to distinguish very small differences in the object being viewed. It directly relates to the aperture of the telescope and the quality of its optics.

**retrograde motion (117)** is when a planet or other body rotates in the reverse direction to what is considered normal rotation.

**right ascension RA (186)** are lines of longitude projected onto the celestial sphere and are measured in hours, minutes and seconds from where the celestial Equator intersects the ecliptic.

**rilles (28)** are long, twisting grooves seen on the surface of maria.

**satellite (14)** is a smaller object orbiting a larger object such as natural satellites or moons.

**seconds of arc (91)** or **arcseconds** is a measurement of angle whereby one degree (of arc) is divided into 60 minutes of arc and each of these into 60 seconds of arc.

**selenology (35)** is the scientific study of the moon.

**semi-diurnal tides (13)** have two highs and low tides per day.

**SERENDIP (235)** is the acronym for Search for Extra-terrestrial Radio Emissions from Nearby Developed Intelligent Populations is a program operating since 1979 which allows for SETI-style searches during the usual on-going routine work at some radio telescopes at the University of California and at the Arecibo observatory in Puerto Rico.

**SETI. (231)** or the Search for Extra-Terrestrial Intelligence, is a series of government and privately-funded radio telescope programs (most of which have been cancelled) designed to detect radio signals from intelligent life from other planets. There has been no positive result so far.

**shephead moons (158) are small moons within the rings of Saturn which seem to maintain the orbits of the smaller particles within the rings.**

**sidereal rotation period (3)** or day is the time for one complete rotation of a planet relative to the background stars.

**sidereal day (96)** is the time taken for one complete rotation of a planet in respect to a distant star. For Earth, this sidereal time is 23 hours, 56 minutes and 4 seconds.

**solar flares (113) are smaller eruptions of material from the Sun's surface.**

**solar prominences (113) are gigantic eruptions** in which large masses are thrown out from the Sun's surface in great magnetic loops.

**solar wind (7)** is the strong flow of charged particles streaming out from the Sun.

**solstices (9)** occur when the rays of the Sun are striking the Earth's surface from directly overhead. They occur in the Northern Hemisphere between December 20th to December 23rd and in the Southern Hemisphere between June 20th and June22nd each year.

**South Celestial Pole (200)** is that imaginary point in the sky of the Southern Hemisphere around which the stars appear to rotate. this is directly overhead of the south geographical pole.

**spectrogram (84)** or spectrograph, is the pattern of bands produced by a spectroscope.

**spectroscope (84)** is a device used to analyse light and uses a prism or grating to disperse the light into its component wavelengths.

**spectroscopy (84)** is the technique which analyses the chemical nature of an object by looking at the light from through a spectroscope.

**spectrum (83)** is a range of wavelengths of electromagnetic radiation such as the visible spectrum of light consisting of the colours of red, orange, yellow, green, blue, indigo and violet – Roy G. Biv is the memory aid.

**speculum (87)** is an alloy of two-thirds copper and one-third tin which can be highly polished and used as a mirror in reflecting telescopes.

**spherical aberration (92)** is a distorted image, especially around the edges of an image, because of the uneven nature of the ground surface of the lens or mirror.

**spring tides (12)** are high tides slightly higher than is usual, so called as the water seems to spring up, not because of the season.

**star (100)** is a large bodies which generate its own radiation, mostly as light and heat, by thermonuclear fusion reactions, unlike planets, moons and asteroids which simply reflect this radiation.

**star clusters (184)** are large but loose groups of stars without any recognizable pattern.

**steady state theory (228)** was a discarded theory of the universe which suggested that it had no beginning but was simply always renewed.

**Sunspots (113)** are the dark, magnetc disturbances of lower temperatures observed on the Sun's surface.

**superior planets (118)** are those planets beyond the orbit of Earth.

**supernova (102)** is the massive explosion which occurs at the end of a very large star's lifetime.

**synchronous rotation (15)** is when one object (e.g. the moon) orbits another at the same time that it rotates therefore presenting the same face to the primary body.

**syzygy (12)** is when the moon and the Sun are in line and produce the greatest tides on Earth such as during spring tides.

**tachocline boundary (106)** is the transition region in the Sun between the radiative interior and the rotating outer zone of convection currents.

**terrae (28)** are the upland region of mountains on the Moon.

**terrestrial planet (126)** is one having a similarity tom Earth in size and having a rocky and metallic interior.

**tidal (gravitational) locking (15)** occurs when the influence of a larger body locks the rotation of an orbiting body so that it rotates at the same time as its orbit e.g. the Moon always has one side facing Earth.

**Titus-Bode law (143)** was an 18th century mathematical sequence which seemed to predict the positions of planets from the Sun. it was later discredited.

**transit (108)** the observed passage of a planet across the face of the Sun.

**Trojans (178)** see **Trojan asteroids.**

**Trojan asteroids (146)** are small rocky bodies which lie in the orbit of Jupiter at the leading (l4) and trailing (l5) Lagrangian points.

**umbra (19)** is the total, darker shadow made when an object is shielded from the Sun during an eclipse.

**tropics (9)** are the places across the Earth north or south of the Equator where the Sun is overhead at mid summer. they are at the parallel of latitude 23°26' north of the Equator as the Tropic of Cancer or south as the Tropic of Capricorn.

**UTC (60) from the** French for *Temps universel coordonné* or Universal Coordinated Time, is the modern scientific time scale used on Earth which was developed from and is almost identical to Greenwich mean time.

**Virgo Supercluster (224)** contains the local group of galaxies and is itself a component of the Laniakea Supercluster.

**vernal equinox (186)** is that point on the celestial Equator conventionally marking the beginning of spring equinox (march) in most Northern Hemisphere cultures and is the prime meridian for right ascension.

**waxing (18)** refers to the phase of the Moon appearing to grow.

**waining (18)** is when the shape of the Moon appears to become thinner.

**white dwarfs (219)** are small stars at the end of their life cycle which simply cool off after exhausting their hydrogen. Much of the helium is converted to carbon which fails to re-ignite and electrons are compressed into the smallest possible space available.

**zenith (9)** is the point in the sky directly overhead of the observer.

**zodiac (9, 188)** refers to the 12 constellations which lie on the ecliptic and through which the Sun, Moon and planets appear to move.

This book is also available in electronic format which can be purchased at amazonz.com for Kindle or other electronic devices such as PCs and iPad using the free Kindle App. Books in the series **ADVENTURES IN EARTH SCIENCE** are available from Felix Publishing, Australia (info@felixpublishing.com) and include:

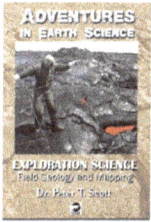

EXPLORATION SCIENCE
Field Geology &
Mapping

FOSSILS – LIFE in the
ROCKS

RICHES from the
EARTH
Minerals & Energy

A DANGEROUS PLANET
Volcanoes &
Earthquakes

CHANGING the
SURFACE
Erosion &
Landscapes

THROUGH SEA and SKY
Oceanography &
Meteorology

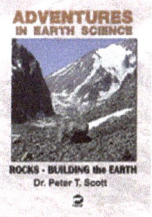

ROCKS – BUILDING
the EARTH

BEYOND PLANET
EARTH
An Introduction
to Astronomy

www.ingramcontent.com/pod-product-compliance
Lightning Source LLC
Chambersburg PA
CBHW060350220326

41598CB00023B/2869